Earthquake Strengthening
for Vulnerable Homes

A practical guide for Engineers, Contractors, Inspectors, and Homeowners

by Thor Matteson, Structural Engineer
author of *Wood-Framed Shear Wall Construction—an Illustrated Guide*

Earthquake Strengthening for Vulnerable Homes

A practical guide for Engineers, Contractors, Inspectors, and Homeowners

ISBN: 978-0-9906645-0-5

Copyright 2014

by
Thor Matteson

ALL RIGHTS RESERVED. This publication is a copyrighted work owned by the author. Without advance written permission from the copyright owner, no part of this book may be reproduced, distributed, or transmitted in any form or by any means, including, without limitation, electronic, optical or mechanical means (by way of example and not limitation, photocopying, or recording by or in an information storage retrieval system, electronic or otherwise). For information on permission to copy material, contact the author via e-mail to thor (at) shearwalls dot com

Information presented in this book is general in nature and is not intended as a substitute for the specific advice of a competent, experienced engineer. The author shall not accept any responsibility for the misuse or misapplication of any of the material herein, nor for the suitability of any information presented.

First Printing: December, 2014

Second Printing: June, 2015

Cover Design: Marney Mutch, Shear Seismic Inc, Victoria, British Columbia

Cover Photos: Left, and top right: Michael Sevy, Rebuilding Together San Francisco. Others: Author

In Memory

The structural engineering and home inspection worlds both lost dedicated and respected members in 2014: Ben Schmid, a structural engineer whose career spanned six decades, and Roger Robinson, one of the first to establish and shape the private home inspection profession. Both of them influenced my career and provided assistance with *Wood-Framed Shear Wall Construction—an Illustrated Guide*. Both of them were exceedingly generous with their time. I did not know either Ben or Roger very well, and that is precisely why I know how much they cared about furthering their professions: they took the time to pass along as much knowledge as they could to me, a relative stranger.

Ben Schmid,
Jan. 12, 1923—Aug. 23, 2014
Structural Engineer

Roger Robinson,
Jan. 3, 1945—Oct. 2, 2014
Home Inspector

Ben and Roger were both exemplary professionals who influenced generations of their colleagues. Ben performed research and invented systems for better protecting homes from earthquake damage. Among many other accomplishments, he presented training on wood construction to the Structural Engineers Association of California. Roger was a founding member of the Golden Gate Chapter of the American Society of Home Inspectors, whose members admire his wisdom and focus on providing prospective home buyers with complete, clear, and concise information regarding the largest purchase of their lives.

If you live in a house, you can thank someone like Ben or Roger for assuring that it is safe. They both cast long shadows that will take many years to fade. Thank you, Ben and Roger, for your encouragement, inspiration, and generosity.

Introduction.

Wood-framed houses are very safe during earthquakes. The houses themselves rarely cause injuries. You are more likely to be injured by a collapsing chimney or falling furniture than by structural failure of the house. Therefore engineers and municipalities have focused on strengthening truly deadly buildings, such as unreinforced masonry, tilt-up concrete, non-ductile concrete frames, and the "soft-story" subset of wood-framed buildings, which have proven to perform badly and led to the majority of earthquake deaths.

Earthquakes continue to destroy houses; in the process they destroy lives, if only figuratively. For most people, their house is their most valuable asset. Homeowners need effective and affordable methods to strengthen their homes and protect their families. Wood-framed home construction is like a forgotten step-child of the engineering profession. Engineers have known the basic strengthening measures for houses for a long time. The principles are assumed to be simple and uninteresting. Couple that with the fact that most engineers earned their college degrees so they would not have to crawl under houses, and finding an engineer who specializes in residential earthquake strengthening can be difficult.

This book is not meant as an engineering study of wood-framed construction. Given the huge variety of sizes, shapes, ages, styles, weights, and histories of houses, researchers will never be able to accurately analyze a wide variety of houses. Even if a researcher's computer model was based on *your* house, the phantom representation living in the computer would not reflect reality. Does the model account for the slightly settled and cracked footings? What about the termite damage, non-standard framing conditions, archaic materials, or hidden conditions inflicted over multiple remodel and repair projects through the life of the house? In my opinion, sophisticated modeling of houses is rather futile; there are just too many conditions to address. Some basic conditions certainly warrant further research. Unfortunately these conditions are generally quite dull compared to other projects competing for research dollars.

Engineering research has far outpaced the training for those who install retrofit systems. Knowing the non-linear response and hysteretic properties of shear walls will not help us if those shear walls are being installed incorrectly. We need to master the basics first: nail along each edge of each panel; do not cut slots in panels to fit them around pipes; and so forth. Engineers who think this is obvious are imagining shear walls built in a perfect workspace. Designing earthquake retrofit work without having a good understanding of the obstacles faced in a typical crawlspace may lead to defective installations. If an engineered design shows plywood panels installed where pipes or other obstacles pass through a wall, but does not show how to work around those obstacles, is it fair to expect that a capable but untrained worker will devise a satisfactory solution?

Even if it were economical for an engineer to design the strengthening system for every house that needs upgrading, there are not enough engineers available to complete this task before the next earthquake. Engineers are busy designing new buildings and the planned repairs and remodels of existing buildings. We cannot reexamine every house ever built in earthquake-prone areas.

The primary need I see in the residential retrofit world is training for specialty contractors*—an army of them is needed to retrofit all the vulnerable homes. To begin answering that need, this book shows many typical vulnerabilities that are not addressed in other references and describes practical methods to strengthen them. Some common retrofitting schemes are obsolete, yet are still in widespread use; I explain why these methods should be abandoned (and provide currently accepted alternatives). The 100 pages or so that this book devotes to "simple" houses needs to be expanded to include hillside homes, soft-stories, and split-level houses, among others. There's plenty to do before the next quake—let's get going!

* A designation for such a specialty would also be good

CONTENTS

Chapter 1 Background Information 1
    1.1 Important Messages to Homeowners 1
    1.2 Why This Book? (And why now?) 2
    1.3 Acknowledgements 4
    1.4 This book is not a Code, Standard, or Guideline 5
    1.5 Respect for Intellectual Property 5
    1.6 Disclaimer 6

Chapter 2 Why Retrofit? And How Much Work Should You Do? 7
    2.1 Wood-framed buildings are generally very safe 7
    2.2 Earthquakes can destroy houses 7
    2.3 Which houses are most at risk? 7
    2.4 How earthquakes damage houses 8
    2.5 Homeowners: Yes, you should worry about earthquakes 11
    2.6 What building code applies to earthquake retrofit work? 13
    2.7 How strong a house do you want? 14
    2.8 What if your house is not "typical"? 17
    2.9 What to expect after an earthquake 18
    2.10 Overly-conservative designs may not help homeowners 19
    2.11 Attorneys, liability, and guarantees 19
    2.12 For simple houses, spend your money on construction rather than calculations 19

Chapter 3 Theory, General Information, Design Considerations 21
    3.1 Shear wall basics 21
    3.2 Retrofit Shear Walls 22
    3.3 Continuous load path 25
    3.4 Conventions, abbreviations, and definitions 26
    3.5 Simplified explanations of advanced theory 27
    3.6 Common-sense design approaches 29

Chapter 4 The Real World: Typical Existing Conditions & Specific Retrofit Requirements . 31
    4.1 Existing Construction and Retrofit peculiarities 31
    4.2 Best materials and Hardware 39

vii

  4.3 Practices & materials to avoid 41

  4.4 Know when to stop engineering and start building 42

  4.5 Retrofit priorities 45

Chapter 5 Designing a Basic Earthquake Retrofit 49

  5.1 Basic physical principles that govern earthquake forces 49

  5.2 Square buildings are easy 50

  5.3 Rectangular houses are easy, too—maybe 50

  5.4 Choose your code 51

  5.5 "One size fits all" solutions often don't fit anything very well 52

  5.6 "Plan Set B"—the inside story from a committee survivor 52

Chapter 6 Detailed Construction Methods 55

  6.1 Jobsite safety 55

  6.2 "Standard" Construction Conditions 56

  6.3 Nonstandard Construction Conditions 89

  6.4 Existing obstructions 95

  6.5 Repair Methods for poorly-installed prior retrofit work 99

  6.6 Special bracing & prefabricated shear walls 105

Chapter 7 Construction details 109

  7.1 Mudsill Connections to Footings 109

  7.2 Cripple wall connections to framing above 124

  7.3 Shear transfer diaphragms 133

  7.4 Collectors 137

  7.5 Plywood/miscellaneous 147

Chapter 8 Retrofit Mistakes 155

  8.1 Mis-installation of stock hardware 155

  8.2 "Home-made" hardware and invented connections 160

  8.3 Expensive retrofit work that provides relatively little protection 164

  8.4 Inferior construction methods 171

  8.5 Pressure-treated lumber: a ticking time-bomb since… 2004? 173

Chapter 9 Special Structural Concerns 187

  9.1 Hillside homes 187

  9.2 Soft- or weak-story construction 193

9.3 Irregular and unusually shaped buildings 200

Chapter 10 Foundation Problems; Cracks, Termites, Etc. 203

10.1 Old concrete 203

10.2 Settlement 203

10.3 Cracks in concrete foundations 206

10.4 Rotated footings 207

10.5 Capped footings 209

10.6 Saddled footings 210

10.7 Footing "curbs" 210

10.8 "Bathtub" basements 211

10.9 Brick foundations 211

10.10 Efflorescence 212

10.11 Termite and dry-rot repairs 212

Chapter 11 Stucco Can Resist Earthquakes in Some Cases 215

11.1 Pre-World War II practices 215

11.2 Line-Wire Stucco 217

11.3 "Self-furred" Stucco 218

11.4 Weep screeds 221

11.5 Stucco identification 221

11.6 Retrofitting stucco 222

11.7 Further information on installing NEW stucco 224

Chapter 12 Other Earthquake Hazards 225

12.1 Masonry chimneys are killers 225

12.2 "Non-structural" earthquake hazards 230

Abbreviations & Glossary 236

Appendix A "Standard Plan A" Strengthening Requirements 241

Appendix B Evaluating Existing Retrofit Installations 243

Appendix C Common Hardware Items 245

Appendix D Responses to Arguments Against Flush-cutting Mudsills 247

Appendix E Problems with "Angle Iron" Connections 249

Appendix F Severe Roof Weakness Present in some pre-1920 Houses 261

Appendix G Excerpts from *Assessment of Damage to Residential Buildings Caused by the Northridge Earthquake** 263

Appendix H Excerpts from *Practical Lessons from the Loma Prieta Earthquake* (National Research Council) 267

Appendix I References and Further Reading 271

Chapter 1 Background Information

The idea of an "earthquake-proof" house is a fantasy. Retrofit methods presented in this book are intended to prevent collapse—not to prevent all damage. Anyone living in a house in an area subject to earthquakes should expect their house to be damaged after a sizable quake. This book presents ways to strengthen typical houses so that they remain habitable after an expected earthquake, using efficient and cost-effective methods.

1.1 Important Messages to Homeowners

The advice and construction methods presented later in this book are general in nature. This section is even more general, but very important to understand.

1.1.1 Prepare now; better yet, *over-prepare* now

Your home may never experience the "design event" that an earthquake retrofit is intended to withstand. But if that earthquake does come, your home gets *one chance* to survive it. Most people would rather spend their money on a fancier kitchen remodel than on a bunch of hardware and plywood under their house that nobody will ever see. But the "extra" money you spend to include earthquake retrofit measures in that remodel could well make the difference between you cleaning up the things that shook out of your kitchen cabinets, or having your house red-tagged and bulldozed—along with the remodeled kitchen.

Take as many retrofitting steps as you can afford now, but don't stop at a minimal level of protection. Strengthening your home should be an ongoing process. The longer a fault waits to unleash an earthquake, the more energy is stored up to shake your house apart. Use this time wisely.

Spending additional retrofit dollars now could save you from major inconvenience after an earthquake. Let's say you retrofit your home to a very basic level and it sustains only "minor damage" in a big quake. That same quake will have destroyed tens of thousands of houses whose owners were not as responsible as you were, and construction resources will mostly be diverted to rebuilding those houses. After the Northridge Earthquake in 1994, some homeowners lived in motels for more than two years while waiting for their homes to get rebuilt. "Minor damage" could become extremely annoying after a couple of years, especially if it's buckled flooring that you trip over three times a day, or cracked tile that renders a shower stall unusable, or a living room ceiling that collapsed under the weight of bricks raining from your chimney. Such reminders may make you wish you'd spent a bit more on your retrofit.

1.1.2 Install as much or as little as you can afford, but *install it well*

The recommendations in this book do not set a set the highest standard for earthquake strengthening. Sadly, many existing faulty retrofits fall far short of even a low standard.

I cannot coerce owners to spend more than they can afford in order to achieve a superb retrofit. However, I do want to make sure that whatever components they install are high-quality materials used in an effective manner. Anything less is almost a complete waste of money.

1.2 *Why This Book? (And why now?)*

Fifteen years ago, after seeing repeated problems in new residential construction, I began work on a book that explains shear wall construction. An earthquake retrofit contractor read my book and inquired about several aspects of shear walls constructed under existing homes. This inquiry led to my most fascinating change in career path so far.

Designing shear walls properly in new construction is a challenge—designing them to fit around all the obstacles under a hundred-year-old house is stimulating!

I am currently serving as a paid subconsultant to FEMA, developing what was hoped to be generic retrofit methods for typical wood-framed houses. Trying to apply complex engineering analysis methods to a wide range of house shapes and styles, built over a range of about 130 years, to resist earthquake forces that vary depending on location is an enormously complex process. To keep the project manageable we have to limit what building types our "cook-book" retrofit methods will address. To assure that the retrofit work will be strong enough for almost all of the buildings almost all of the time, the strengthening we recommend will be much stronger than is needed for most of the buildings most of the time. Including construction details that apply to conditions found only occasionally makes the resulting plan set seem more complex than it really is, but if you do not include such details then the plan set will not address that group of houses. (If you omit five details, each of which is needed by only five percent of existing homes, you have potentially reduced the group of homeowners who can use the plan by 25%.)

A much more ambitious project funded partly through FEMA and partly through the California Earthquake Authority is hoped to provide comprehensive retrofit guidelines that will apply to almost all types of homes found in California (which would likely also include most houses on the west coast). Why then am I publishing this book on my own? The main reasons follow.

Increased sense of urgency

In 2009 I began revising and updating "***Wood-Framed Shear Wall Construction—an Illustrated Guide***". On January 12, 2010 a devastating earthquake struck Haiti. A few months later the biggest quake in recorded history (depending on who you believe) hit Chile. Then one jolted Southern California/Northern Baja California. I could not help thinking: *they are getting closer.* How long until a densely populated area of California experienced another large quake, damaging houses that I walk by every day?

Committees are not known for speedy action, and I feel that the public needs access to at least one structural engineer's retrofit recommendations as soon as possible.

In 2003 the US Geological Survey estimated that a damaging earthquake has a 67 percent chance of occurring in the San Francisco Bay Area in the following 30 years. (Similar estimates are available for other areas of the US, and other agencies provide estimates in other countries.) As one geologist points out, the USGS prediction does not mean 29 earthquake-free years and then a 67 percent chance of a quake in the last year.

Advice that official codes cannot give you

The building code, and committees working to develop the code, cannot recommend specific brands of hardware. As an individual I can tell you the exact catalog numbers of useful retrofit hardware. There are also some "generic" products that perform better than others. For instance, both plywood and OSB meet the building code requirements for "wood structural panels"—but I recommend only plywood for strengthening your home.

Code development committees consist of dedicated professionals who volunteer their time and expertise on a particular subject. The individuals in these groups have varying opinions on the subject, which is the best way to produce a balanced code. Any document produced by consensus of a committee will be a compromise. This book is a compromise of a different sort: I have traded thorough review and vetting by a formal group of my peers for getting this book published quickly. If I can provide useful information even a month earlier than a committee of experts could provide a model code, that month might be what gives **you** the time to retrofit *your* house before the next earthquake destroys it. Recommendations in this book are also compromises between quality and affordability, *as is the current building code that governs brand new construction.*

Advice against some practices allowed in the current building code

I recommend against some materials and practices that comply with the building code. For instance, the code allows chemical treatments that prevent termite and fungal damage to wood, but result in severe fastener corrosion (discussed further in Section 8.5).

More timely updates

FEMA has released several publications on earthquake strengthening. The government hires a consulting engineering firm to write such documents. Once the document is written, reviewed, and published, it is nearly impossible to update or revise it; another contract has to be negotiated with the same or a new consultant and a multi-year writing/editing/comment/review cycle completed before final publication. Through direct correspondence, FEMA has acknowledged that corrections need to be made to a couple of their documents, and also that this may never happen.

Awareness of current misunderstandings in the retrofit world

Members of the Golden Gate Chapter of the American Society of Home Inspectors (GGASHI) have provided many photos of earthquake retrofit work. Some of the installations they see will be effective in reducing earthquake damage, but many of them seem to demonstrate poor understanding of retrofit needs. In an informal, non-scientific poll of GGASHI members in September, 2012, the general feeling was that only about 20% of the retrofit installations they saw would protect against "The Big One." (Estimates ranged from 5% to 30%.) The GGASHI members are "in the trenches" every day, and indicate that the state of retrofit practice is not what it should be.

Occasionally I see earthquake retrofit designs prepared by other engineers. Some designs are very well conceived. However, some are so conservative that it is hard to think that an owner could afford them—the common result being that the owner does nothing at all. A retrofit design is useless if it is so expensive that you cannot implement it. Other designs focus on exacting calculations, but omit construction details that show how to strengthen the house. In one case, 10 pages of calculations accompanied a small floor plan with notes indicating where to install plywood bracing panels; *no* construction details were included.

Need for practical details

"Realistic details that are simple and are easily accomplished by tradespeople with a wide range of skill levels will be the most successful in achieving the structural integrity goals set forth in the Preface to the Uniform Building Code." —Rick Tanis, American Plywood Association, May 3, 1990 memorandum to the State of Washington Association of Building Officials regarding lateral strength of buildings.

One main goal of this book is to present a wide variety of practical details. When I wrote *"Wood-Framed Shear Wall Construction—an Illustrated Guide"* ("the **Shear Wall Guide**"— see Reference #1 in Appendix I) one of my goals was to show real-life shear wall installations. There were many references at the time that showed how to build a simplified shear wall. Currently there are several resources for earthquake retrofit methods that present the same details shown in the International Existing Building Code Appendix Chapter A-3, with minor embellishment or modification. This book addresses many conditions that I have not seen presented previously. But no reference can cover all possible construction conditions—that's why engineers are still around.

1.3 Acknowledgements

Thanks to all who provided inspiration and input for this book, and in particular:

GGASHI: The Golden Gate chapter of the American Society of Home Inspectors; truly the upper echelon of professional home inspectors, who provided photographs from hundreds of earthquake retrofit installations, discussion, and thought-provoking questions.

Structural engineers **Nels Roselund** and **Ben Schmid,** for sharing their decades of knowledge and experience.

Engineers and researchers over time and space for their work generating much of the "raw data" in various reports, codes, and other documents, that is distilled and presented here.

Various Contractors who have shared insights with me.

1.4 This book is not a Code, Standard, or Guideline

Unlike most building codes, this book is not a "consensus document." Such a document is in the conception stages; as of 2014 I am serving on a "steering committee" to guide development of a "pre-standard" that will eventually lead to a standard, which will evolve into a code. Judging by the time-scales involved in releasing similar codes, it may take 10 to 15 years to develop a consensus retrofit code that addresses what is covered in this book. If you wait for a consensus document, bear in mind that it will include concepts and methods presented in this book, and that earthquake faults are not going to wait for engineers to come to consensus regarding the many different conditions found under old houses.

A "standard of practice" (SOP) is different from a building code. An SOP is a formal document composed by a professional society or organization. Such a standard related to earthquake retrofits would likely take years to develop—that is, if a professional society decided to develop such a document; many other activities are more important.

A "standard of care" is vaguely related to an SOP, except that a standard of care changes from one project to another. A standard of care is *"the watchfulness, attention, caution and prudence that a similarly situated professional would exercise under the circumstances."* Defining a standard of care is essentially done by a court of law, and the standard will vary depending on "circumstances." Which circumstances? Who knows—the weather, project schedule, budget, available materials, whether the homeowner will allow filling in window openings with plywood, and myriad other factors. Of these, the budget is a big consideration. If a homeowner tells an engineer to spare no expense in a retrofit, the standard of care would be much higher than if they could spend only $8,000 and wanted the most effective retrofit for the money.

1.5 Respect for Intellectual Property

This book represents a great deal of my time and professional expertise and is essentially a donation to the public good. Book sales will never repay my time and expenses. Conversely, the information provided may save your house from complete destruction by an earthquake. All I ask is that you pay the meager publication price and do not redistribute information on your website, in seminars, classes, etc. I hope you will consider this more than a fair exchange. If you are interested in speeding the production of the next edition you are welcome to donate more than the purchase price. You can find me at www.shearwalls.com .

1.6 *Disclaimer*

Information presented in this book is general in nature and is not intended as a substitute for the specific advice of a competent, experienced engineer. The author, publisher, editor, illustrator, printer, typesetter, book binder, and contributors shall not accept any responsibility for the misuse or misapplication of any of the material herein, nor for the suitability of any details presented.

Specific mention of various hardware products is not a guarantee that they will perform in any particular application during an earthquake or any other event.

Use of photos or illustrations provided by others shall not be taken as an endorsement of the photographer or illustrator.

This book presents the author's views, his interpretations of research performed by others, and solutions to common installation conditions. Solutions, recommendations, and opinions presented do not necessarily reflect those of any of the contributors, nor any professional society to which the author belongs, nor any entity that has used the author as a paid or volunteer consultant.

Chapter 2 Why Retrofit? And How Much Work Should You Do?

2.1 *Wood-framed buildings are generally very safe*

Your wood-framed house will probably not kill or injure you in an earthquake. Furniture and other things in your house are more likely to injure you than the house itself. But most of us cannot afford to lose our home. Earthquake insurance can help you rebuild, but you will have to pay out a very substantial portion of the costs before the insurance company begins to pay.

2.2 *Earthquakes can destroy houses*

Wood-framed houses, especially those built before 1960, can suffer serious damage in earthquakes. If you are holding this book and read this far, I probably don't need to show you lots of photos of earthquake devastation. Figure 2-1 shows two convincing photos.

Figure 2-1 Some typical earthquake damage from the 1994 Northridge earthquake. FEMA photos

2.3 *Which houses are most at risk?*

Houses with "crawlspaces" below the first floor are generally more vulnerable to earthquakes than houses built on concrete slab foundations (also called "slab-on-grade" or "slab-on-ground" foundations). If you are not sure what sort of foundation system your house has, you can get some clues by looking at the outside walls of your house.

If you see screened vents or an access hatch outside your house as shown in Figure 2-2 then you definitely have a crawlspace. For houses with very low crawlspaces the floor framing is usually supported directly on the foundation sill, or "**mudsill.**" The mudsill is the lowest piece of wood framing that bears on the concrete or masonry foundation. Taller crawlspaces often have "**cripple walls**" built on top of the mudsill; these are short wood-framed walls that extend up to the bottom of the first floor framing. In some regions it is common for tall concrete or masonry foundation walls to support the floor framing directly.

Figure 2-2   Vent openings in outside walls indicate the house has a crawlspace.

You can often tell whether you have a crawlspace (and what sort) by counting the steps to your front or back door. One step (or no step at all) means you probably have a slab foundation. Two or three steps mean you probably have a low crawlspace with no cripple walls. Three or more steps indicate a taller crawlspace and likely presence of cripple walls.

2.4 How earthquakes damage houses

The headline read, "*Most earthquake damage is caused by shaking*". I can't think of any earthquake damage that is *not* caused by shaking, directly or indirectly. To complete this thought it needs to be noted that **most earthquake damage is caused by shaking** *from side to side***.**

2.4.1 Houses do *not* jump up off their foundations!

The biggest misunderstanding of earthquakes (based on poor retrofits I see in the field) is that the major concern is houses lifting up off their foundations. Engineers talk about "hold-downs" or "tie-downs" that are used to resist "uplift" forces. ***Tie-downs do not keep your house from lifting off the foundation:*** they keep *part* of your house from lifting up—namely, one end of a shear wall. The other end of the shear wall is exerting an equal force down on your foundation. Hurricanes and high winds can generate pure uplift forces that require different hardware than earthquakes, yet many, many earthquake retrofits include various types of "hurricane ties" that are ineffective in resisting side-to-side forces.

> **Hold-downs installed by themselves are useless:** *if no plywood bracing panels are installed, no force can transfer to a hold-down or to the foundation.*

2.4.2 Weaknesses to look for

The following categories of structural weaknesses are described in FEMA Publication P-593, "*Seismic Rehabilitation Training for One and Two-Family Dwellings.*" Some houses may have problems not listed. The first two conditions are the most common hazards and generally lead to the greatest damage in earthquakes; these are the primary focus of this book.

1. **Inadequate connection from the footings to the "mudsill:"**
   The mudsill is the wood member that connects wood house framing above to the concrete or masonry foundation. Until the 1937 building code, no fasteners were specified for this critical connection.
   Between 1937 and 1997, the building code required bolting the mudsill to the foundation with ½-inch diameter anchor bolts spaced no more than 6 feet apart. Since 1997 the code has required bigger bolts (5/8-inch diameter) at closer spacing (no more than 4 feet apart).

2. **Lack of cripple-wall bracing:**
   Cripple-walls are the short wood-framed walls that support the main floor of your house above the foundation. Until the 1980s these walls were often too weak to adequately resist earthquake forces. Materials and construction quality have declined since the 1970s to the point that the cripple walls in newer houses often need strengthening. If the cripple-walls collapse, portions of the house will fall several feet to the ground. This usually results in extreme damage and leaves your home uninhabitable.

3. **Heavy brick, stone, or other chimneys:**
   Masonry chimneys can be very dangerous. They are usually very expensive to brace.

   A falling chimney caused the only death in the 1992 Landers earthquake.

   One school of thought regarding chimneys is that removing them and patching the void left in the house would cost about as much as repairing damage if they were to collapse during an earthquake.

   Avoiding costs of possible future damage-repair is one thing; preventing injury or loss of life is another. If you choose not to remove the chimney, try to spend as little time as possible where it could fall. Locate favorite arm-chairs, sofas, desks, sewing and hobby tables, etc. as far as practical from the fireplace or chimney. Children should not have sleep-overs near the chimney.

   ABAG advises placing plywood or planking on top of the ceiling framing around the chimney to help stop bricks from falling into the room below during an earthquake. This would work until bricks pile up on top of the plywood and collapse the ceiling and might give you just enough time to get away from the chimney.

4. **Open front buildings:** True open-front buildings are usually older multi-unit apartments built above commercial spaces with large store-front windows, or newer apartments built above "tuck-under" parking. This weakness is very similar to "weak wall lines" under Item 8 below. (This is often referred to as a "soft-story" condition, which is not entirely precise; other construction can lead to soft-story conditions, such as un-braced cripple walls.)

5. **Split-levels:** When two segments of a house have different floor levels without sufficient connections between them, the segments can separate and collapse. An example of a split-level floor system is a house with living, dining and kitchen areas over a low crawlspace, with stairs that lead up half a story to bedrooms above the garage. Depending on the original construction methods used, this split level in combination with the open front at the garage can be particularly hazardous.

6. **Archaic pre-1920 roof framing:** Somewhat rare even in older houses; see Appendix F.

7. **Hillside homes:** Houses built on significant slopes perform poorly in earthquakes. Houses built on the downhill side of the street generally fare worse than those uphill from the street. Special measures are often needed to tie main-level floor framing to the highest portion of the foundation. Ideally in a retrofit the floor framing can connect directly to the foundation rather than relying on indirect connections through walls of different heights.

> *"Hillside homes pose, without question, the greatest life safety risk of any type of damaged dwelling investigated by this subcommittee."*
> —Comment from report on damage caused to wood-framed homes by the 1994 Northridge Earthquake (had they investigated unreinforced masonry homes, hillside homes would have been only the second-most dangerous)

8. **Weak wall lines:** Houses where one or more sides do not have enough solid wall length to resist earthquake forces perform badly in earthquakes. Often found in row-houses in San Francisco, where the entire front wall of one or more floors consists of a large garage door opening, or perhaps a garage door opening with adjacent access doors. Weak wall lines are often found in the front and rear walls of houses on narrow lots and houses with large window or door openings that dominate a wall or walls. Buildings with great views to the Bay, Sound, ocean, or elsewhere often have weak wall lines. This weakness is very similar to an open-front building.

9. **Post connections:** Where posts provide the only support for segments of a home, secure connections at the top and bottom of each post are essential. [If perimeter walls surround many short interior posts, this is a very different situation; in this case it is best to focus on bracing around the perimeter of the house. Untold retrofit dollars have been wasted supplementing short posts, as discussed in Section 8.3.1]

10. **Outside stairs, decks, canopies:** The main concern with exterior stairs, porch roofs, or canopies is that they will collapse and trap you inside the house. Weak construction above exits poses additional danger if people try to exit the building during an earthquake (which is not a good thing to do).

11. **Masonry veneer:** Brick or stone "veneer" (facing or cladding that is attached to wood-framed walls) increases earthquake hazards for two reasons: it can fall off and injure or kill people, and it increases the earthquake forces on a building. Brick or stone veneer less than four feet above ground level is generally not considered a serious hazard, although if it is near a building exit you may wish to remove it or securely attach it to the building.

The FEMA guidelines also include unbraced water heaters, which deserve serious discussion in a book—but not this one. In addition to the concerns in the FEMA list is the following:

12. **Foundation weaknesses**. This category includes the following:

- Deteriorated foundations that are no longer suitable for connecting the wood structure above.
- Stone masonry that would make connections difficult or unreliable.
- "Post-and-pier" construction, where the perimeter walls are supported on isolated posts instead of a continuous foundation. (*Post and piers inside a crawlspace surrounded by a perimeter foundation almost never need supplemental connections*).

2.4.3 Strengthening below the first floor is almost always most important

For most old houses the most vulnerable part of the structure is below the first framed floor level. There are three main reasons for this: First, in most houses only the perimeter walls have sheathing or siding that extends down from the first floor level to the foundation. Crawlspaces are usually open areas with no bracing at all besides the perimeter walls. Above the first floor, the interior walls between rooms provide much more resistance in comparison. Second, only one side of the cripple walls has siding or sheathing applied. Even though it is weak and brittle in comparison to plywood, interior lath and plaster adds considerable strength to exterior walls above the first floor, as does gypsum wall board in more modern homes.

The third reason to focus on cripple walls is that they must resist more earthquake force than walls above the first floor. Earthquake loading depends on the amount of weight shaking back and forth above a particular level. The very lowest part of a structure must resist the greatest force.

In short: Compared to walls above the first floor level, crawlspace walls are less extensive, weaker, and have to resist more earthquake force. Strengthening those areas is the "low-hanging fruit" for earthquake retrofits.

2.5 Homeowners: Yes, you should worry about earthquakes

I hope you are holding this book because you already want to strengthen your home. In case you are not yet convinced, let me address the most common fantasies that people subscribe to.

"My house made it through the last earthquake with no damage"

Most of my work is in the San Francisco Bay Area. An amazing number of people think that the 1989 Loma Prieta Earthquake, that was centered about 60 miles away, compares to a quake that might occur 60 yards away. The argument that "my house survived the Loma Prieta/ Northridge/Nisqually/fill-in-the-blank earthquake" might be valid if your house was *right at the epicenter* of that quake. Otherwise you may as well add that your house survived Hurricane Katrina, the 2011 tornado season, the 1991 Oakland Hills Fire, and the eruption of Mount Saint Helens. Don't think you're close to a fault?

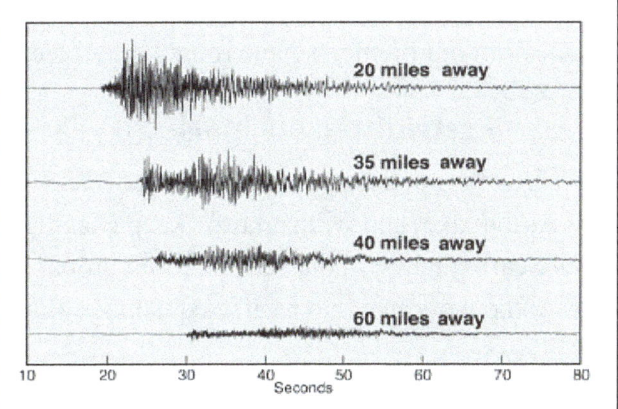

Figure 2-3 This image from the US Geological Survey shows the difference in shaking intensity from the same earthquake recorded at stations varying from 20 to 60 miles from the epicenter. Many homes in California are located much closer to active faults.

Many recent earthquakes occurred on faults that were previously unknown—including the 1987 Whittier Narrows earthquake, 1994 Northridge quake, 2000 Napa quake, and 2003 San Simeon quake.

"My house has withstood the test of time"

All buildings eventually *fail* the test of time. Time has allowed termites, carpenter ants, powder-post beetles, wood bees, and decay to weaken your structure. Tree roots have grown under and heaved the foundation. Gophers have burrowed back and forth under your footings and caused them to settle. Remodeling projects took out sections of walls to open the floor plan or add windows or doors. You replaced the light-weight wood shingles on the roof with new plywood overlain with concrete tile roofing.

Earthquake faults are very patient as we are making our buildings heavier and weaker. In any one area, it may be centuries between earthquakes; ask the people in L'Aquila, Italy, some of whose buildings had stood for 900 years until a quake destroyed them in 2009. Older nations commonly experience earthquakes that destroy 400-year-old buildings, if for no other reason than earthquakes don't occur very often. An earthquake in 1812 leveled New Madrid in Missouri; if there had been more than 200 non-native settlers in the area at the time, this would have been a much more remembered event in our euro-centric history.

"My house was built to code"

Codes keep changing, for a couple of reasons: they don't include all the requirements needed for good earthquake-resistant construction (see the last paragraphs of Section 4.1.2.3); and we keep learning more about how buildings react to earthquakes. Over the years, the foundation anchors required in active earthquake areas have increased in size and quantity; the current connection requirements have roughly doubled over the last 20 years.

"I retrofitted my house 20 years ago"

Best retrofit practices continue to improve. State of the art has changed significantly since the last earthquake, and will certainly keep changing as we learn more about what works best in future earthquakes. Your retrofit could probably use at least a "tune-up." As mentioned earlier the work may have suffered damage from decay or pests, or tradesworkers with chainsaws.

Looking through hardware manufacturer's catalogs and books written before the 1989 Loma Prieta Earthquake describing how to prepare for earthquakes, you find many outdated and even denounced practices.

"The previous owner retrofitted my house 5 years ago"

If the previous owner was an engineer with extensive retrofit experience, then good. But one of the prime reasons for writing this book is seeing many recent retrofits that offer almost no earthquake protection. How complete was the retrofit? What materials and hardware were used? One of the worst was a retrofit installation done in 2012. I was called to evaluate it in early 2013, and had to tell the owner that the work they had paid for was essentially worthless.

"My house is built on bedrock"

If you are within a few miles of a fault, it really does not matter if you are on rock or soil—earthquake forces on your house will be about the same. If you are farther away from the earthquake source the forces are magnified by soft soils; this does not mean they are zero for rocky sites! In California, many rocky areas are also hilly; a house on a sloping site is far more vulnerable in an earthquake than one on flat terrain.

"My house was designed by Frank Lloyd Wright/ Bernard Maybeck/Samuel Maclure/another Famous Architect"

All the more reason to retrofit your house, then! Even though great architects of the past designed exquisite houses, they didn't know much more about earthquake-resistant design than others of the time.

2.6 What building code applies to earthquake retrofit work?

If you apply for a building permit for a project that has "earthquake" or "seismic" in the description, you have two choices. The International Existing Building Code (**IEBC**) or the International Building Code (**IBC**). [In California, substitute a "C" for the "I;" we work with the California Building Code.] The IEBC Appendix Chapter A3 applies to houses that meet certain limits (the most notable and problematic being that cripple wall height cannot exceed 4 feet). The IEBC gives prescriptive requirements for strengthening the structure below the first floor. In cases where engineering is required, components can be designed considering only 75% of the forces required under the current building code. The IBC applies to work that cannot be done under the IEBC. If you are applying for reduced permit fees or other incentives that some jurisdictions offer, they may have specific requirements you must follow.

Leaving the words "earthquake" or "seismic" out of your permit application and plans offers you more freedom. Chapter 34 of the IBC applies to existing buildings. You can make any structural alterations you wish to an existing building as long as you do not make it any weaker than its current condition. Calling a project a "voluntary strengthening" or the like allows you to do as much or as little strengthening as you can afford. One problem with this approach is that the official permit record leaves only ambiguous clues as to what work was done and how effective it might be. The sword cuts both ways, and many ineffective retrofits have been installed that, while they don't make the building worse off, do much less to improve things than the owners had probably hoped.

2.7 How strong a house do you want?

If you are building a new house, I recommend spending an extra 5 to 8 percent and making the earthquake and wind resisting components about twice as strong as the code requires. This additional strength would greatly reduce damage during an earthquake.

If your house is already built, the better question is, "How strong can you afford to make your house?" For most people I suggest you spend as much as you think you can afford, plus a little more. And then make further improvements in a few years—and so on.

In my professional practice, my designs follow the IEBC to the extent that I generally limit work to the crawlspace (see Section 2.4.3). I usually design for higher loads than required in the IEBC. As a licensed professional I can also design for conditions that the IEBC does not recognize, since I can apply the correct engineering principles.

The foregoing approach is what most homeowners can afford. Of course you can choose to go beyond basic strengthening measures, up to the point of taking your house apart and rebuilding it piece by piece to meet the current building code or beyond.

Can you over-do an earthquake retrofit?

Initially you might think that stronger is better; but the increased cost beyond what the IEBC considers "strong enough" under your house would probably be better spent in strengthening first-floor walls around large window openings, or implementing other strengthening measures. In the worst case, the construction bids you get will be so high that you can't afford to do anything. Having an unused set of plans for a super-strong retrofit won't protect you during the next quake! Later sections discuss common ways where, in my opinion, designers can go far beyond what is reasonable in their requirements.

There is also ongoing debate between engineers whether strengthening the cripple walls beyond a certain degree could actually lead to more damage to the floors above. This argument should be settled shortly after the next big quake, but until then, as a practical matter it is hard to get enough plywood bracing under most houses to make them "too strong."

2.7.1 Goal for earthquake performance:

Let's set up a strength scale of 1 to 10, where "1" is a building that would fall down in a moderate quake, and "10" represents a building that will survive a very strong quake and hardly even suffer cosmetic damage. In the middle of the scale, say "5," we have buildings that are strong enough to survive a major earthquake, but they suffer so much damage that they fall down in an aftershock.

Many people are surprised to learn that new construction meeting the current building code gives you a "5" on the above scale. That's right: if you build a house that meets the code, it may have to be completely rebuilt after "The Big One." In reality, nothing is built to level

"10;" not even hospitals, nuclear power plants, or emergency services buildings (though the latter are very close).

Trying to bring an existing house up to a "5" can be very expensive—especially if you need to address weak areas above the first floor. Using the methods presented in this book can get you to level 4 or above *for the crawlspace or basement areas of your house only*. Retrofit work above the first floor level is beyond the scope of this book. Whether you decide to upgrade the substructure of your house to a 4, 5, or 8 depends on your level of paranoia and how much money you want to spend. The amount of hardware and other reinforcing determines the strength of the retrofit. My recommendation is generally to install as much bracing plywood as room allows without moving furnace ducts, water and gas lines, electric panels, waterheaters and the like. Sometimes you just have to move such obstructions. Sometimes you need to replace the foundation under part of your house in order to get to a level 4. Maybe you can afford that, or maybe getting to a "3" in that particular location will make you feel safe enough. Let the next person who buys the house rebuild that particular foundation section.

Few people can afford a retrofit that will let their house come through an earthquake with little or no damage—but most houses can be strengthened to survive without *major* damage. Don't worry about getting an A+ on an exam when you need only a C-minus or even a D.

Some engineers do not feel comfortable designing to meet less than what the building code would require for new construction. Some engineers even go beyond that. If you work with an engineer, make sure that their "comfort level" for the design matches your own comfort level, but more importantly that it matches your *budget*.

Earthquake forces are the least at the roof level, and unless you have solid brick walls (which means this book will not help you) then it is almost inconceivable that roof connections will affect a basic retrofit design. (Note: Some older houses may have roof connections that are in danger of serious damage during earthquakes. See Appendix F for an explanation of this rare condition.)

2.7.2 "Strength is essential—but otherwise unimportant" -*Hardy Cross*

Almost every structural engineer will tell you that it is more important to have a complete structural system than it is to worry about calculating the "exact" forces that an earthquake could produce or designing the structure to "exactly" match those forces. Calculating the earthquake loads is only a minor part of designing an effective earthquake retrofit system. Knowing how to connect all the necessary building components together—and conveying those requirements to the builder—is essential.

Three main factors go into calculating earthquake forces on a building: the expected amount of shaking intensity an earthquake would produce, the ductility of the structure (see Section 3.5.2), and the weight of the building materials. For an older house with many unknown structural components, only the third factor can be estimated with some defensible accuracy.

The first two factors depend on many best estimates. Many significant earthquakes occurred on faults that were unknown before the earthquake (most notably the 1994 Northridge Earthquake). Engineers can estimate how a house will respond to an earthquake, but we should not fool ourselves into thinking that these estimates will always (or possibly ever) be accurate.

2.7.3 Making a "resilient" community

Community planners talk more and more about "resiliency" after a disaster. A resilient community returns to normal quickly after a disaster. If you can stay in your house, keep your job, help with earthquake recovery for everyone else, the whole community is better off. Likewise if all your neighbors can stay and keep working as grocery clerks, bus drivers, teachers, bank tellers, and so forth, then your life will be better as well. After Hurricane Katrina, the city of New Orleans changed forever; preparing for disasters can prevent major, negative regional changes.

Moderate earthquakes in the magnitude 6 to 7 range cause the most damage overall. They shake strongly enough to do significant damage, and they are much more frequent than quakes greater than 7. The message from engineers, geologists, and planners working toward resiliency is to strengthen existing buildings to resist the 6 to 7 earthquakes; new buildings should be designed and constructed to survive the 7 and stronger events.

In her informative lectures, Dr. Lucille Jones of the USGS states that fires will double the damage inflicted by a major earthquake, and that lost business revenue will double the total again. Thus, *preventing earthquake damage reduces overall losses by a factor of four*.

2.7.4 Minimum goal: Your house should be habitable after an earthquake, or "Safe Enough to Stay"

You don't want to live in a tent because your house slid off its foundation or the cripple walls collapsed. You may not have running water, electricity or heat, but at least you have a place to stay. A carefully installed under-floor retrofit is intended to allow you to stay in your house after an earthquake. The house will likely suffer damage—broken glass, cracked or fallen plaster, shifted door-frames and so forth—but the structure will be largely intact. You want your house to pass the earthquake "test" with a letter-grade of *D* or above. Each step above this basic level of protection becomes increasingly expensive.

There is no way for anyone to guarantee that a retrofit will keep your house from collapsing. Geologists and engineers keep refining their understanding of earthquake performance.

2.7.5 Spend your money on work you will never see

The most cost-effective strengthening work is usually done in the crawlspace or basement area of your house, as outlined in Section 2.4.3. For "typical" houses on level or gently sloping lots, doing retrofit work in the living space is rarely justifiable, and almost never necessary to give you the D-minus level of protection.

Part of the value of your home is the character of the neighborhood. What if your house is intact after an earthquake but all the neighboring houses are torn down and replaced with faux-Craftsman schlock or panelized, factory-built imitation Victorian mansions? Instead of paying hundreds of thousands to "earthquake-proof" your own house—and I say this in all seriousness—paying for your neighbors to retrofit their cripple walls may be a better investment.

2.8 *What if your house is not "typical"?*

So far we have talked about "typical" houses. Houses with any of the following features usually need special attention from an engineer experienced in retrofit designs for wood-framed houses.

Soft, Weak, or Open Front ("SWOF", also called "weak-story" or "soft-story") The most common example is a garage with living space above, where there is very little wall length on either side of the garage door. See Figure 2-4.

Figure 2-4 The large garage door openings in the left photo create a "weak story" condition in the building. The FEMA photo above right shows a collapsed weak story—the red arrows show where the roof of the single-story portion of the house used to connect to the collapsed portion, where you can see the "ghost" roof outline.

Masonry homes: Unreinforced masonry buildings (URMs) are extremely dangerous in earthquakes. See Figure 2-5. Retrofits for URMs should be designed by only experienced engineers. An effective earthquake retrofit for an unreinforced masonry house could easily cost a hundred thousand dollars or more. As I will discuss later, just having a brick chimney is dangerous—you don't want to be surrounded by unstable walls.

Figure 2-5 These people were lucky to escape from this unreinforced brick house.

Hillside homes: Houses built on slopes present serious retrofit challenges. Houses on the downhill side of the street ("downslope" houses) are generally worse off than houses on the uphill side ("upslope" houses). Figure 2-6 shows two downslope houses, and the remains of a downslope house after an earthquake.

Figure 2-6 Hillside houses are extremely vulnerable in earthquakes. The right hand photo shows the remains of a hillside house after the Loma Prieta earthquake. Photos: Left—John Fryer, GGASHI Center—Author Right—J. K. Nakata, FEMA

2.9 *What to expect after an earthquake*

Even new structures will be significantly damaged by an expected earthquake; the building codes set a very low bar for protecting property. You cannot reasonably retrofit an existing building to meet current building code requirements—especially if the building was constructed before any codes were in place.

The retrofit methods presented in this book are intended to protect your house from collapse. Depending on what degree of strengthening measures you can install it may be reasonable to give additional protection to your house. For most retrofits, though, the expected earthquake will cause serious damage. Engineers talk about "structural" and "non-structural" damage. Often the non-structural, or cosmetic damage, is most expensive to repair.

"Cosmetic damage" often results in the total loss of a vehicle after a car accident; a side-swipe that scrapes the entire length of a car almost always costs more to repair than the car is worth—yet the car may still be safe to drive. Just as cosmetic surgery is expensive, cosmetic repairs to a house can be very expensive. Cosmetic repairs could include replacing all the exterior stucco, or tile work, plaster, cabinetry and trim, and other components that account for much more cost than the structural frame of a house.

For new buildings, engineers can double the strength of a building for a five to eight percent increase in the building's overall cost. This can reduce cosmetic damage to almost unnoticeable levels for most earthquakes. However, doubling an existing building's strength is extremely costly. It may be less expensive to tear it down and start over.

2.10 *Overly-conservative designs may not help homeowners*

Besides protecting your safety, engineering is about balancing cost versus benefit. Part of this balance is matching the strength of all the components in your house's structural system. In my view it does not make sense to strengthen your floor system to withstand a cataclysmic earthquake in which the rest of your house collapses into rubble on top of the undamaged floor.

2.11 *Attorneys, liability, and guarantees*

Behind every overly-conservative engineer there is likely an attorney (or the specter of one). NO retrofit will guarantee that your building will be undamaged in an earthquake or that you will not sustain injuries. Insurance companies that issue professional liability insurance for engineers will exclude coverage for any sort of "guarantee;" don't even think of asking for one, especially if you want to spend less than 10% of the value of your home on a retrofit design and installation.

2.12 *For simple houses, spend your money on construction rather than calculations*

The idea behind ABAG's *Standard Plan A* (also known as "Plan Set A"), Los Angeles' *Standard Plan 1,* Seattle's *Project Impact* and similar retrofit guidelines is for homeowners and contractors to perform retrofit work without specific engineering. I fully support making it more affordable to retrofit houses using standard details—as long as the houses themselves are "standard."

The purpose of this book is to explain more fully the principles behind effective retrofit construction, in hopes that more of the "standard" houses can be retrofitted without adding a few thousand dollars in engineering costs. An engineered design can easily increase the cost of a retrofit by 20 to 30 percent. If the increased cost puts the retrofit project out of reach of a homeowner, then I would suggest they retrofit their house without engineering—as long as good retrofit methods are used.

Chapter 3 Theory, General Information, Design Considerations

This chapter introduces the basic engineering concepts that apply to wood-framed buildings. We will look at how different building components such as shear walls are supposed to work, but not at specifics of building a shear wall—that comes later. The following chapters will cover specific connections, hardware items, how to select and install mudsill anchors, etc.

If you have studied geology, physics, or engineering, you learned that earthquakes move the ground under a building and the ground exerts a force on the building. This is only partially true—the instant the ground stops moving, the house exerts a force on the ground until movement is damped out.

Most people find it easier to think that earthquake forces travel from the house into the ground. For the purposes of installing retrofit components and designing a retrofit, it makes absolutely no difference which direction the forces are traveling; therefore in this book I will refer to earthquake forces originating in the building and traveling to the ground. Physicists may care where the forces originate; but their houses don't, and yours won't either.

3.1 Shear wall basics

This book is an expansion of Chapter 6 in the second edition of *"Wood-Framed Shear Wall Construction—an Illustrated Guide"* (hereafter referred to as "the **Shear Wall Guide**"—see Reference #1 in Appendix I). The Shear Wall Guide is about 180 pages long, with over 150 photos, diagrams, and illustrations, and explains the principles and construction shear walls in considerable detail. There is no way to present all the applicable information in this book. Anyone performing an earthquake retrofit or building a new wood-framed house should study the *Shear Wall Guide*. Only the very basic principles of shear walls are repeated here.

In this book "shear wall" means plywood sheathing attached to wood framing in a way that will resist earthquake forces that act from side to side. "Braced wall," "plywood bracing panels," "bracing panels" and similar terms are used to mean the same thing.

The most basic shear wall principle is that each of the four edges of every piece of plywood used in the shear wall must be connected to a framing member (stud or end-post, mudsill, top plate, block, or other member as appropriate). Depending on which framing member we are considering, the connections either transfer loads into the plywood, or transfer loads from the plywood to the framing.

3.2 Retrofit Shear Walls

Adding proper connections and bracing (plywood) to a cripple wall turns it into a shear wall. The following principles apply; for in-depth information, refer to the *Shear Wall Guide*.

3.2.1 Action—Shear Transfer Ties deliver loads to the top of the wall

Forces have to act on the shear wall if it is going to do any good. We are assuming earthquake forces originate in the house and move through the shear walls to the foundation and into the ground. The cripple wall top plates deliver force to the plywood shear panels. Before this can occur, though, we need to connect the floor framing that is supported on the cripple wall to the top plates.

The action force travels through either existing fasteners or new connectors between the floor joists and the cripple wall. The generic term I will use for *new* connectors is "shear transfer tie," or **STT** for short. In houses without cripple walls, STTs are installed from the floor framing directly to the mudsill.

> **Shear transfer ties may even be needed in *new* construction.** Building codes have progressed toward requiring a secure connection between the floor framing members and the structure below them. However there is still not a clear requirement for connecting floor framing to a mudsill. (This was pointed out by an observant contractor about 20 years after major changes to the building code were implemented to assure a complete load path.) A code change to take care of this oversight will go into effect in the 2015 IBC. Until then, houses can be built that *immediately* need retrofitting under FEMA and other guidelines. This is discussed further in Section 4.1.2.3.2.

3.2.2 Reaction—Mudsill anchors keep the wall from sliding

The *reaction* force occurs along the bottom of the wall. The shear panels connect to the mudsill; the mudsill must connect securely to the footing in order to transfer the earthquake force from the cripple wall into the foundation.

The reaction force is provided by mudsill anchors such as bolts or special foundation anchor plates. Even if there is no cripple-wall you still need to anchor the mudsill to the foundation.

3.2.3 Uplift (Rocking)—Weight of the building (or special hardware) keeps the wall from rocking

When you apply a force along the top edge of a shear wall, one end of the wall tends to lift up off of the foundation. "Tie-downs" or "hold-downs" may be needed to keep the ends of the shear wall from rocking up and down as the earthquake forces shakes back and forth at the top of the wall. The amount of uplift depends on the height of the shear wall, amount of earthquake force, and dead load that bears on the wall.

The term "hold-down" has apparently led to great misunderstanding about what forces we need to resist. Earthquakes do not make houses jump up off of their foundations. We do not need to hold down a shear wall—we need to hold down *one end* of the wall at a time. Since earthquake forces cycle back and forth, uplift forces have to be resisted at both ends—but not at the same time.

Existing retrofits often have tie-down hardware as the only component installed. **Hold-downs installed by themselves are useless**—if no shear panels are installed, no force can get transmitted to a tie-down. Section 8.1 shows many examples of ineffective retrofits with tie-down hardware installed with no regard for side-to-side forces. Figure 3-1 shows an example of why tie-down hardware may not be effective at all in older foundations—they simply do not have enough strength or weight to resist much uplift force.

> *We do not need to hold down the shear wall—we need to hold down one end of the wall at a time.*

Under the current building codes that govern "conventional light-frame construction," uplift is ignored in houses with one story above a crawlspace for braced-walls up to *ten feet tall*, as long as they are at least four feet wide. Requirements were even less restrictive up until about 2006. Uplift for cripple walls less than *four* feet tall is ignored in the IEBC and other retrofit guidelines.

Figure 3-1 This 100-year-old footing was exposed during foundation replacement work. It is only about 6 inches deep, with a total concrete height of 10 inches. With no steel reinforcing bars this footing has very little strength to resist uplift forces.

I would never recommend building a new house with 10-foot tall, 4-foot wide braced walls without engineering—but retrofitting similar houses that are already occupied by people with limited budgets should not have more restrictive requirements than new construction. Providing uplift resistance in an existing building can be very expensive. Since the goal of this book is to explain the most cost-effective earthquake retrofits, we will often not consider uplift. There are few reports of uplift failures causing significant damage to single family homes in recent earthquakes.

In many cases the weight of the building above will keep the cripple wall from rocking up and down. Tie-downs may be prudent for walls that are taller than they are wide, especially if they are parallel to the floor framing.

3.2.4 Bearing capacity—Posts or double studs provide strength at the ends of braced walls

Like uplift, bearing capacity is important at only one end of a shear wall at a time. The end posts must have adequate support on framing below (usually the mudsill). Providing this support is usually not difficult for cripple wall retrofits. Mudsill crushing could degrade the performance of a shear wall; but like uplift, short walls are less likely than tall walls to have problems with crushing.

3.2.5 Internal Strength—Blocking and proper panel connections create a continuous structural unit

The best shear wall would be a single sheet of plywood installed full-length of the building. Since you cannot buy plywood in 30-foot long sheets, much less drag that big a piece of plywood under your house and muscle it into place, you need to join several panels together to make a shear wall.

Repeating the "basic shear wall principle," we need to fasten each edge of plywood that makes up our shear wall. Where panels butt into one another, they need to join on a common member (a single stud, a post, or studs fastened together to act as a common member).

We can't expect shear panels to perform well if they are slotted to fit over pipes, or if we chop them out to install piping or wiring in the future.

Plywood provides bracing strength to the cripple wall based on thickness and quality of plywood, size and spacing of nails, quality and thickness of existing wall framing. Existing construction often includes many obstacles to shear panel installation: wiring, conduit, water and gas lines, sewer pipes, ducts, beams, gas meters, breaker panels, windows, and so forth. Notching, slotting, bending plywood to fit around existing obstacles can seriously reduce the shear wall's strength—so can a future tradesworker who needs to install or repair utilities behind our plywood.

3.2.6 Putting it all together

Figure 3-2 shows how a shear wall fails when the preceding requirements are not met. [Note: this is not meant to be an engineering text book, so only the imaginary applied force at the top of the shear wall is shown.] If the structure above the shear wall is not connected to the top of the wall, it slides off and leaves the shear wall intact; if the shear wall does not connect to the foundation, the shear wall and structure above slide off the foundation together; if the shear wall does not have adequate strength, it breaks apart; if the horizontal force at the top of the shear wall causes the structure to overturn (and this tendency is not resisted by tie-down anchors at the ends of the shear walls) then the ends of the shear wall can lift off of the foundation; if the support for the shear wall is not adequate to support the wall, it can cause crushing.

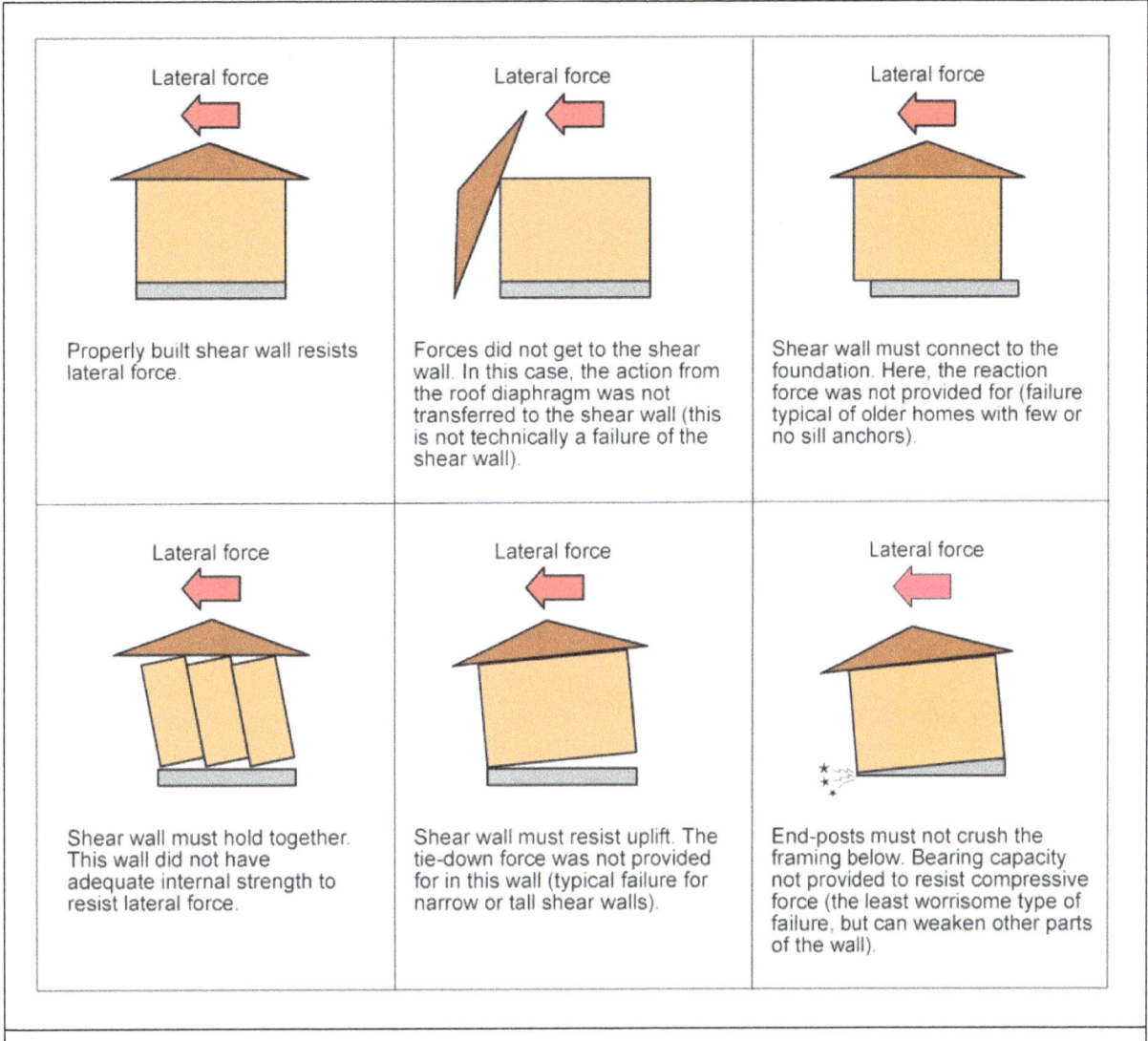

Figure 3-2 Shear wall can fail in the five ways shown.

3.3 Continuous load path

Components of a house must tie together well enough to transmit earthquake forces from the point where they originate down to the foundation and into the ground. If structural members do not provide a complete load path, then other parts of the house (plaster, trim, doors, etc.) try to resist loads. If the "non-structural" elements are strong enough—and they often are—then that portion of your house will survive an earthquake without undue damage. At places where a break occurs in the load path, you can expect the structure to break during an earthquake.

The load path is not easy to visualize. It may be different depending on which direction the earthquake force is exerted. For horizontal forces, anywhere that building components are supported on a horizontal surface below them you need to have a secure connection between the components. Such locations include: Foundation to mudsill, mudsill to cripple studs, cripple studs to cripple wall top plate, top plate to floor framing members, floor framing members to floor sheathing, floor sheathing to walls above, and so on.

Uplift or compressive forces also need continuous load paths. Sometimes the weight of the structure above can resist uplift forces—in this case the load path dissipates into upper floors, walls, and roof framing. When uplift forces are greater than the weight of the building components above, then anchorage to the foundation and ultimately the underlying soil is required. For the purposes of a cripple-wall retrofit the load path is quite direct; for new, multistory construction it can be very complex.

3.4 Conventions, abbreviations, and definitions

3.4.1 General comments

The majority of my experience crawling under houses has been in the San Francisco Bay Area. Regional construction methods vary. For example, some areas often have full basements; in some places it was common to apply siding diagonally rather than horizontally. Almost all of the photos presented in this book were taken under houses in the greater SF Bay Area, with guest appearances by houses in other west coast regions.

Framing practices in the Bay Area vary considerably, and cover a wide variety of conditions. I expect that construction found elsewhere will at least be similar to some of the conditions encountered in the Bay Area—though with differing frequency.

3.4.2 Conventions and language used in this book

Hardware items are most easily and specifically referred to using the manufacturer's catalog number, such as Simpson Strong Tie Company's "UFP10". Groups of numbers and capital letters (sometimes used with quotation marks) designate a specific hardware item. Nails are discussed in great detail in the Shear Wall Guide. Sizes of nails used to be given in "penny" designations, which corresponded to a given length, diameter, and head size. Unless stated otherwise, penny designations refer to "common" nails for diameter and length.

Relative dimensions of the width and height of shear walls are important. Unfortunately, "short" is the opposite of both "long" and "tall." When comparing horizontal dimensions I try to use "wide" and "narrow" as opposites. To compare heights I use "tall" and "short."

3.4.3 Common terms used

Figure 3-3 illustrates most of the common components in a cripple wall. See the Glossary for other terms and expanded definitions.

Note that this figure shows only one of several possible ways to construct a cripple wall.

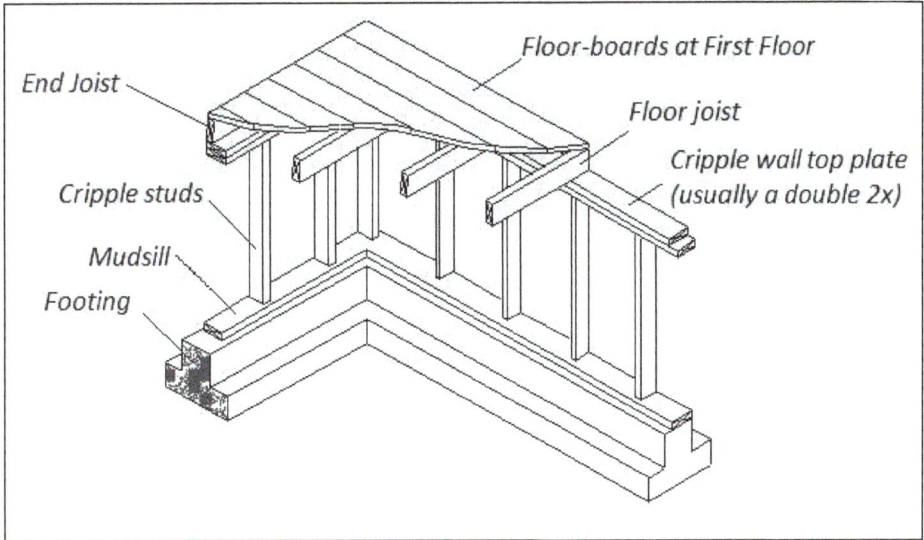

Figure 3-3 Typical structural components found in retrofit work. For clarity, blocking between joists, exterior cripple wall finish, walls above first floor, soil, and other non-structural items are not shown. Illustration courtesy of Bay Area Retrofit, Inc.

3.5 Simplified explanations of advanced theory

Various structural materials and systems can behave very differently. The following explanations are intended to give non-engineers at least a basic understanding of some important concepts that apply to earthquake-resistant design.

3.5.1 "Cyclic loading" versus "Static loading"

Earthquakes shake back and forth (and also up and down, which is much less important to us). A significant earthquake can produce dozens or even hundreds of back and forth cycles. Each loading cycle degrades the strength of a structural system. Material tests using cyclic loading give the best prediction of how a component or system will perform in an earthquake.

"Static loading," on the other hand, shows only how much load something can resist as a force is applied in one direction and increased gradually until failure. A greater factor of safety is usually applied if something has been tested under only static loading—that is to say, we trust static load tests less than cyclic load tests.

3.5.2 *Ductile* versus *brittle* materials or behavior

"Ductility" is a measure of how much a structural material, component, or system can deform without failing, and how many load cycles the system can endure. Ductility also measures how suddenly a system fails. Ductile materials and systems are preferred in earthquake resistant design because they will resist many loading cycles.

Metal components are typically ductile. Steel can be formulated for amazing ductility. Steel reinforcing bars can provide ductility to concrete and masonry systems.

In contrast, brittle materials include brick, unreinforced concrete, plaster, and stucco. These systems may have great strength up until the point of failure—but they usually fail suddenly and without much warning.

Ductile materials absorb earthquake energy. In a wood-framed shear wall the ductile portion of the system is the panel edge-nailing. As the earthquake shakes to and fro, each nail bends back and forth and absorbs some of the earthquake energy. The more energy a system can absorb without failing, and the less energy is transferred into the rest of the building, the better.

One good example of ductile versus brittle behavior is the way a metal teaspoon will bend almost double back and forth, compared to a plastic cafeteria spoon that bends somewhat before breaking suddenly.

3.5.3 Relative stiffness

Stiff, or rigid, structures or elements do not deform much under load when compared to flexible structures or elements. Mixing rigid and flexible systems or components can change how forces are resisted—often in very unexpected ways.

A material's stiffness is not the same as its strength: a rope is strong and flexible; a brick is strong and brittle. Steel and wood are "flexible" compared to concrete, plaster, and masonry.

When an earthquake shakes a building, forces are distributed to various parts of the building based on their *stiffness* relative to the other parts. If the *strength* of a particular element is not sufficient to resist the imposed force, then it will fail. One case where this is important to consider is when shear walls are constructed on a stepped footing. Tall shear wall segments are much more flexible than short shear wall segments; the short segments fail first, followed by failures of the successively taller wall segments (this is discussed in greater detail in Chapter 4 of the *Shear Wall Guide*). Graduated shear wall heights on stepped foundations are common in hill-side homes, and tend to fail because the entire length of shear wall does not act together to resist sideways forces.

3.5.4 Building response to earthquake motion

In the same earthquake, adjacent buildings may perform very differently. Assuming the same soil conditions under the buildings, their behavior depends on the stiffness of each building, its ductility, height, weight, and strength.

The earthquake force that an engineer calculates for a particular building is based partly on the "response factor." The response factor is given the term "R" in the "base shear equation. The response factor depends mainly on the stiffness (rigidity) and the ductility of the structural system.

Plywood shear wall systems are much more ductile than stucco, lath and plaster, or gypsum wallboard. The response factor, R, for plywood shear walls allows engineers to design for much less earthquake force than required for an otherwise identical building that used stucco, plaster, or gypsum board to resist earthquake forces.

3.6 *Common-sense design approaches*

Curiosity and fear are sometimes considered the best emotions for engineers to have. Curiosity makes you consider all the ways a connection or structural member could fail; fear makes you think it would fail in the first place. When you need to stretch retrofit dollars as far as possible, I would add practicality as an equally important trait.

Retrofit designers are often faced with compromises between a really expensive solution needed to meet "the code" (or other requirements) and keeping the project within the owner's budget so they can actually go through with construction. A good question to ask in such cases is, *"What would be the consequence of this member or connection failing?"*

If a failure could result in loss of support for a whole wing of a house, then I pay more attention to that compared to a failure that might result in only localized damage. For example, post-to-beam connections may have different importance depending on what the post supports. Posts supporting a small area of floor are much less critical than posts that support a whole room above. Figure 3-4 shows examples of "expendable" posts versus crucial ones. (For more on post connections, see Section 8.3.)

I also consider what interaction there might be between different retrofit components, and what is even a "failure." Retrofit mudsill anchors are an example; expansion anchors are one common choice for this connection. If the anchors you choose are supposed to be tightened to a certain

Figure 3-4 *Top:* Someone went to a lot of trouble to sandwich all the stubby posts between plywood as shown in the top photo. If these had been left unbraced and any of them came loose, other adjacent posts would provide support and limit damage.
Bottom: This post supports an entire building segment; keeping it in place is essential, but can be done with very simple and inexpensive hardware.
Top photo: Author
Bottom: Michael Brady, GGASHI

torque, but start to spin in the hole at only half that amount, how big a problem do you have? Expansion anchors should be used only in shear anyway, and they would resist shear quite well if you didn't tighten them at all. If overturning is resisted by tie-downs or the weight of the structure above, your retrofit may perform adequately with "loose" anchors.

The foregoing is not meant as forgiveness for sloppy work or poor materials, but as an acknowledgment that sometimes pursuit of perfection clouds the view of doing an adequate and affordable job.

Chapter 4 The Real World: Typical Existing Conditions & Specific Retrofit Requirements

This chapter explains some of the common construction found in old houses, material recommendations, and suggested priorities for retrofit work.

The following is still somewhat general in nature; specific connections are discussed in greater detail in Chapter 6.

4.1 Existing Construction and Retrofit peculiarities

For carpenters and engineers accustomed to working only with contemporary tools and methods, some aspects of century-old construction will be unfamiliar. Wood buildings older than 150 years were usually timber framed, but not many buildings on the west coast are that old. Most wood-framed buildings are either platform or balloon framed. This book does not address construction methods such as mortise-and-tenon frames.

4.1.1 Quality of materials

The lumber available through the 1960's, before we cut down all the old-growth trees, was amazingly free of defects. Old lumber has tight grain, few knots or splits, and is usually straight and true. Before 1940 or so, lumber was often left rough. A 2x4 was often a full 2" x 4" or at least 1¾" x 3¾" (vs. the current 1½" x 3½"). Together, the higher quality and thicker boards make it harder to split old lumber when nailing into it. The photos in Figure 4-1 show examples of wood quality from 2012 vs. 1960.

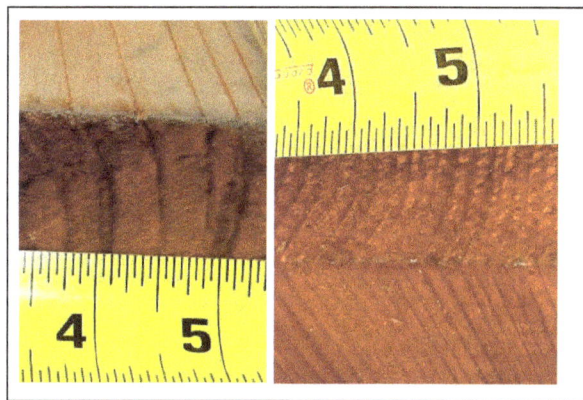

Figure 4-1 Compare the growth rings of old lumber vs. new: *Left*—lumber delivered in 2012. *Right*—a floor joist from 1960. Closer ring spacing generally means stronger lumber

Concrete quality varies wildly. Testing by the Structural Engineers Association of Southern California indicated that for mudsill connections to old concrete, failure of the concrete never occurred before the wood failed. I have seen concrete worse than what they tested, though. Another structural engineer has suggested that if concrete does not break when you drill into it to set retrofit anchors then it's probably strong enough for retrofit connections. Figure 4-2 shows some examples of old concrete.

Figure 4-2 Old concrete. *Top left*: Photographed in 2012, this 102-year-old sidewalk looks like it has a couple of centuries of life left.
Top right: As long as you don't bolt directly to the "honeycombed" area, this concrete would likely be fine for supporting a retrofit. Photo by Paul Barraza, GGASHI
Lower left: This footing may be strong enough to provide support during an earthquake, but in another decade or two it would not. Replace the footing or anchor shear walls to other portions that are in better condition. Photo by Paul Barraza, GGASHI.
Lower right: This footing should be replaced. Photo by Roger Robinson, GGASHI

Concrete conditions in the same foundation system can vary widely. If the original concrete was mixed by hand in wheelbarrows or small drum mixers on site, or in different weather, by different workers using materials of different quality, then the original foundation had variations throughout. Over the lifetime of the foundation some areas may have deteriorated more than others. The only way to get a really good idea of concrete strength is to take a core sample to a testing laboratory. This is almost never done. Concrete can be tested in place by using a calibrated device (known as a "Schmidt hammer," after the inventor) that measures how far a spring-loaded piston rebounds after striking the concrete. Either of these methods gives you a result only at the location of the test, so you may need several tests if your concrete quality changes much from one area to another.

A Schmidt hammer is not something that many contractors or homeowners are likely to go out and buy. Someday FEMA or other authorities may devise a simple way to test existing concrete. Until then it is rather subjective—when you hit it with a hammer you should get a good rebound and ringing, not a dull thud. If you can break out individual pieces of aggregate, that's a bad sign. Cracks and settlement do not necessarily indicate poor concrete quality, but may be evidence of other problems that require foundation work. (See Section 10.3).

Brick foundations also vary in quality. When foundations were still commonly built from brick, lime mortar (with no Portland cement) was typically used. Over time, lime mortar loses its strength when exposed to air (exposure to water makes it degrade even more quickly). Many deteriorating brick foundations were covered with a layer of plaster, or capped or "saddled" with concrete.

While there is lively disagreement about the adequacy of brick foundations to support retrofit connections, I am not aware of brick *foundation* failures that lead to collapse of the structure. Note that data from the August 24, 2014 South Napa Earthquake is still coming in—but I doubt that foundation failures played a major part in the collapse of any wood-framed buildings. See Section 10.9.

Once you install earthquake retrofit work it becomes much more difficult to replace the foundation. If the foundation looks questionable it is best to replace it before doing any earthquake retrofit work, or as part of the retrofit. Figure 4-3 illustrates a practical reason to make foundation repairs before the retrofit—foundation contractors are not always respectful of retrofit work that is in their way. See the sidebar for further thoughts on replacement.

Figure 4-3 The plywood bracing panel was cut out to provide access for a foundation contractor to pour a new footing. Too bad they did not replace the plywood—the installation was probably stronger with intact plywood and poor original concrete. Photo by John McComas, GGASHI

When should you replace your foundation? If it looks really ugly, chances are it will look ugly to the next person who wants to buy your house. If you replace it now then there is no room for argument that it's "not good enough." And trying to work with old concrete or brick, or put "bench piers" under failing sections of foundation can be almost as expensive as just replacing it. One option is to replace the worst portions and let future owners do the rest. If the concrete or brick is in questionable condition now, what will it be like by the time an earthquake hits?

4.1.2 Assumptions about original construction

Houses that need retrofit work may have been constructed using plain boards and hand tools, or using plywood and modern power tools. The cost of materials used to be much higher compared to the cost of labor. Understanding some of the old construction methods can help in deciding what is important to strengthen.

> **Brick foundations** have a bad reputation that is not always justified. Thousands of unreinforced brick buildings have been strengthened to resist earthquakes. **Most of those retrofitted brick buildings** *have brick foundations.*
> The earthquake forces from a brick building are about ten times as great as from a wood building—if a brick foundation is good enough for a brick building, *in general* it should not be viewed as unacceptable in a wood building.

4.1.2.1 Floor framing methods

Floor joists need to be stabilized so they do not roll over. Current practice is to butt the joists against a "rim joist" and nail through the rim joist into the ends of the regular joists. The end-nails prevent the joists from flopping over until plywood subflooring is attached to them, providing final stability. The current codes usually require blocking between joists at intermediate supports such as beams or interior bearing walls.

When labor was less expensive it was common to use smaller floor joists and connect them at mid-span with "cross-bridging" to help spread loads between adjacent joists. The cross-bridging also kept the joists from rolling over, and blocking was often not installed above intermediate supports as is required now.

Prior to using rim-joists, three methods were commonly used to stabilize joists at exterior walls: Blocking between all joists; blocking in every other joist bay; no blocking at all (stability provided by wood siding or wall sheathing). Sometimes the blocking is not the same depth as the joists.

The various framing methods have their advantages and drawbacks, which are discussed in detail later.

4.1.2.2 Perimeter nailing at floor diaphragm

Most original conditions that need retrofit work are visible from underneath a house. In some cases the exact connection is concealed, but adding new connectors is relatively easy to accomplish. One notably difficult connection to either verify or retrofit is between the original floor sheathing and the floor joists. The IEBC does not require supplementing this connection. Some engineers do not want to rely on a connection they cannot see. This often leads to expensive, likely unnecessary retrofit connections such as the ones shown in Figure 4-4. The following explains why such supplementary connections are seldom needed.

Plain lumber sheathing (individual boards, rather than plywood) was commonly used on floors up until the 1950s. Since then there was a trend toward tongue-and-groove plank subfloors and eventually plywood and OSB. The transitions occurred at different times depending on material availability and local traditions.

Plain lumber sheathing generally consists of one-inch nominal boards nailed to the floor framing. Sheathing boards were installed perpendicular to the joists ("straight sheathing") or at an angle—usually 45 degrees—to the joists ("diagonal sheathing"). Since time immemorial, two 8-penny common nails were used at each joist to connect boards up to 8 inches wide, and three nails were used to connect boards from 8 to 12 inches wide.

Figure 4-4 *Left*—The two "A35" connectors at the top of the photo add great expense compared to strength provided.
Center—This photo was taken looking straight up at the underside of the floor. An "A35" connector is fastened to the underside of the floor sheathing boards with short screws (so the points of the screws do not poke up through the floor). Short screws have reduced load capacity, so you need more of them to resist a given load. The original sheathing boards are 4 inches wide, and each board is almost certainly fastened to the joist with two nails. Contractors often charge between $25 and $40 for each A35 installed like this; specifying them every 4 feet around a house's perimeter can add $1,000 to $1,500 to the cost of a retrofit, only to strengthen a connection that is already adequate.
Right—Revenge of the Retrofitter; after dropping all seven screws onto his face and losing them in the powdery crawlspace soil, the installer gave up and left this A35 unconnected. This photo shows diagonal sheathing, and the original construction provides a very sturdy floor diaphragm without further connections.
Photos: Left—author; Center & Right—Paul Rude, GGASHI

Floor sheathing boards were spliced as needed along the length of the building. Good practice was to stagger the splices (as current good practice is to stagger plywood joints). The quickest way to install sheathing boards was to let them "run wild" past the outline of the floor, and then once they were all nailed in place make a single cut along the edge of the floor (either using a hand saw or a power saw). Cutting around the edge of the floor platform would have been difficult if the sheathing was not firmly attached to the floor joists or blocking below; it would have been especially challenging using a hand saw, as the boards would chatter up and down with each stroke of the saw. (My grandfather was a carpenter for a time, and it is hard to imagine him or any other carpenter omitting this connection even if he was drunk while working late on a rainy Friday evening.)

Forces in the floor diaphragm are spread out fairly uniformly. The nailing described above would consist of 8-penny common nails spaced at four to six inches. This does not count nailing from subsequent construction through the flooring and into the framing, such as at wall sole plates. In my opinion calling for added connectors between the original floor sheathing and framing members is almost never warranted. This opinion is supported by Section A304.1.3 of the CEBC or IEBC which states in part that along rim joists or blocking *"the existing top edge connection need not be verified."*

4.1.2.3 Toe-nails from joists to mudsill (or to cripple wall top-plate)

The connection from the floor framing members to the cripple-walls or mudsill below is a very important part of the load path. There are two common conditions found in every house: where the joists are perpendicular to the cripple wall or footing below, and where the joists are parallel to the wall or footing. The first case is often adequate; the latter condition usually needs strengthening (*even the current code does not call for adequate connection in all situations—See Section 4.1.2.3.2*)

4.1.2.3.1 Joist connections at bearing points

Floor joists are supported at each end by a bearing wall, beam, mudsill, or other structural member (in some cases the floor overhangs the support, but the connection is the same). Most pre-1950 houses have at least two toe-nails from joists to the supporting framing, as shown in Figure 4-5. The older the construction, the larger the nails (typically). Houses built before 1930 often had 20-penny nails at this connection, and 16-penny nails were typical up until the 1950s. Current codes require connecting joists to framing below with three 8-penny toe-nails.

The toe-nail connections are often concealed by blocking between the joists, but it is extremely rare for them to be completely missing; the connections were essential in keeping the framing from falling down as the carpenters built it. For joists spaced at 16 inches on center (by far the most common spacing in the Bay Area) the toe-nail connections provide a shear resistance of about 100 pounds per foot of wall length. This is almost always enough capacity for single-story homes, unless they feature heavy construction such as thick stucco or Spanish tile roofs.

Figure 4-5 Typical toe-nailed connection from joist to cripple wall top plate. (Note that there is no block in this joist bay, but the nail at the top of the photo fastens a block on the far side of the joist.)

If you really need to save money on a retrofit installation, STTs *along bearing walls* or supports are much less important than other components, especially in a long, narrow house. There are two reasons that shear transfer ties on the long walls become a lower priority in a long house. First, the floor joists almost always bear on the long exterior walls in such houses. As mentioned above, this gives an existing shear capacity of about 100 pounds per foot. (Contrast this to the narrow walls, where a joist runs parallel to the cripple wall and may have a nail only every few feet.) We can be fairly certain that there are more nails per foot of wall length along the long walls than along the narrow walls. The second reason is simply that the long walls are just that—long. Each foot of wall length gives you more strength to resist forces. This connection can almost always be strengthened later with little additional trouble.

Conversely, even **new** construction may need STTs along joists that are parallel to supporting members, as explained in the next section.

4.1.2.3.2 Joist connections along parallel supports

Until 1994 there was no code requirement for the crucial connection between an end-joist and the cripple-wall or mudsill below. Common practice was to provide nails every four feet or so along the joist; just enough to hold it straight until the floor sheathing was installed and provided more stability. This connection may need to resist several thousand pounds of earthquake force; a few toe-nails proved much too weak in the Loma Prieta earthquake.

Since 1994 the UBC and subsequent codes have required that "rim joists" be connected to the "top plate of the wall below" with 8-penny toe-nails spaced at 6 inches or less. This increases by five times or more the strength of the connection from joist to cripple-wall top plate *when the nails are really installed.*

There are two problems that persist in the current code: First, the term "rim joist" is not defined anywhere in the code; second, there is no requirement for connecting from joists to *mudsills* in cases where there is no cripple wall. To some carpenters, a rim joist means the member that all the regular joists butt into. The joist that bears along the length of a cripple wall or mudsill is usually called an *end joist*. One could argue that three toe-nails from the end-joist to the cripple wall top plate meet the code requirement. The bigger concern is the complete lack of requirement to connect from any sort of floor framing member to a mudsill, in the common case where floor framing bears directly on the mudsill.

4.1.2.4 Pairs of 30-penny spikes from mudsill into footing

Mudsill anchor bolts were not required by code until 1937. California did not have consistent code adoption or enforcement until well after that; houses built before 1960 may not have anchor bolts. Before anchor bolts were required carpenters used other methods to secure mudsills to footings. One common practice was to drive 20-penny or 30-penny spikes part way into pieces of mudsill and then turn the pieces over and embed the exposed heads of the spikes into freshly-poured concrete. Figure 4-6 shows two such spikes after sections of mudsill were removed from the top of the footing. Spikes were typically used in pairs spaced about 24 inches apart along the mudsills. This connection method completely conceals the fasteners, but is very common. After all, carpenters had to pound lots more nails into the mudsills and cripple studs, and did not want anything moving around while they did so.

Figure 4-6 *Left*: The nail shown was embedded in the original mudsill, which has just been cut back flush with the stud face and partially removed. The nail measured around 0.21 inches in diameter. *Right*: Similar construction across the San Francisco Bay (viewed from below, where a chunk of foundation concrete had broken off)

Another method of attachment was to drive spikes all the way through the pieces of mudsill, and embed the points of the spikes in the fresh concrete. These connections are visible when not covered by cripple-studs or decades worth of dust and debris.

While I usually recommend adding mudsill connections (even if there are existing anchor bolts) it is quite possible that mudsill connections will survive without supplementation. If you cannot afford to install mudsill connections along with other retrofit components, go ahead with other retrofit work; install additional mudsill connections as soon as you can. (You should NOT rely on the original connections forever! Additional mudsill anchors can be installed to either side of shear panels, or UFP10s can be installed over the plywood.)

4.1.3 Termite and pest damage
(and "repairs" that butcher structural elements)

Even the best-quality lumber is subject to pest damage. So-called "dry" rot or other decay is extremely insidious; by the time a trained wood specialist can detect decay using a microscope, wood may have lost 50% of its strength. (Search for *"wood decay pick-test"* on the web for info on low-tech decay investigation; or as of June 2015 www.inspectorsjournal.com/forum/topic.asp?TOPIC_ID=2101&SearchTerms=pick+test)

Attaching new plywood to rotten or damaged framing will give little earthquake protection. Likewise, replacing damaged members without addressing the cause of damage (usually water entry) will offer very short-lived protection. Proper repair methods are far beyond the scope of this book.

Pest extermination contractors are experts in identifying and killing fungi and bugs. They are not trained in structural principles, and rarely tie new members into the original structural system. If pest repairs have been made, additional connections may be needed to provide for a continuous load path.

4.2 Best materials and Hardware

Most of the cost of a retrofit is labor. Spending $25 less to buy low-quality materials degrades the quality of a retrofit by far more than $25. In some cases the better products actually cost less. I currently specify the following materials and hardware.

- Borate-treated lumber for mudsills. Section 8.5 addresses severe corrosion problems with other lumber treatments. Borate-treated lumber is usually less expensive than lumber treated with other chemicals. California Cascade Industries, 800-339-6480 supplies borate-treated lumber throughout California, often with only a few days lead time. Many lumber yards will tell you that what they have is just fine. Do not believe them. If you special-order borate-treated lumber, ***check to make sure*** that it really is. There will be a plastic tag identifying it with "DOT" or "SBX" on the end, or sometimes this information will appear stamped on the boards.
- Five-ply, structural 1 grade plywood, 15/32" thick (usually referred to as "half-inch" thick) as shown in Figure 4-7 is the best material for wood shear walls. The additional layers of wood reduce the chances that splits or other defects will align from one layer to another—possibly inside the panel where you cannot see them.

- Lumber should be Number 1 grade Douglas Fir. This grade has fewer and smaller knots and other defects than Standard & Better or Number 2.
- Use "engineered lumber" such as laminated veneer lumber (LVL) and plywood where appropriate. Wood shrinks and splits as it dries. Split structural members cannot resist loads as effectively as sound members; perhaps not at all. Figure 4-7 shows a block that shrank about ¼" as it slowly dried out after installation under a building; the resulting split makes it worthless for its intended purpose. LVL and plywood will not shrink and split like ordinary sawn boards. All of the existing lumber in an old house has reached its equilibrium moisture content and will not shrink any further. New lumber that you install needs to remain the same size to perform well. This is an especially important consideration for blocking, because short lengths of lumber split easily—and blocks are usually installed because you need to transfer shear forces from the top to the bottom of the block.

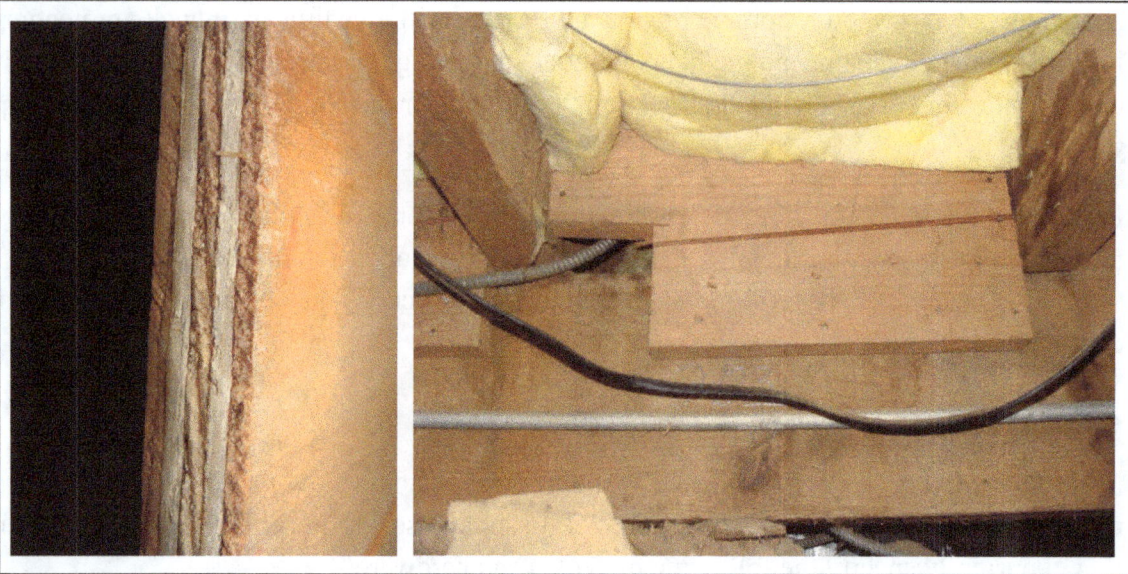

Figure 4-7 Good retrofits depend on good materials.
Left: **GOOD: Five-ply, 1/2" plywood;**
Right: **BAD: Lumber that shrinks. The block shown at the top of the photo was installed between 130-year old joists that had all dried and shrunk. The top and bottom edges of the block were fastened tightly to the original framing; when the new block dried and shrank, it split and lost any usefulness. To avoid this, use kiln-dried lumber, or, better yet, LVL or other "engineered lumber."**

- Some framing connectors are much easier to install than others. Simpson's L series (L50 & L90 are examples) have one leg that is 2-3/8" long, compared to the 1½" long legs on their A34 & A35 connectors. (For KC Metals, the CA50 and CA90 have longer legs; for USP, use the AC5 & AC9). The longer leg allows for better nailing access when you are reaching up into an awkward space with bulky gloves and tools. Install the long leg of the connector so the nails will not split a sliver of wood off of the connected members—See Figure 4-8. These may be installed with 8-penny x 1-1/2" nails if 10 penny nails tend to split the wood.
- "Plywood nails" for attaching shear plywood are shorter than "common" nails. Only 1½" nail embedment is required into studs for plywood attachment. This means that nails longer than 2 1/8" are not only unnecessary for plywood attachment, but can be harmful if they tend to split the existing framing.

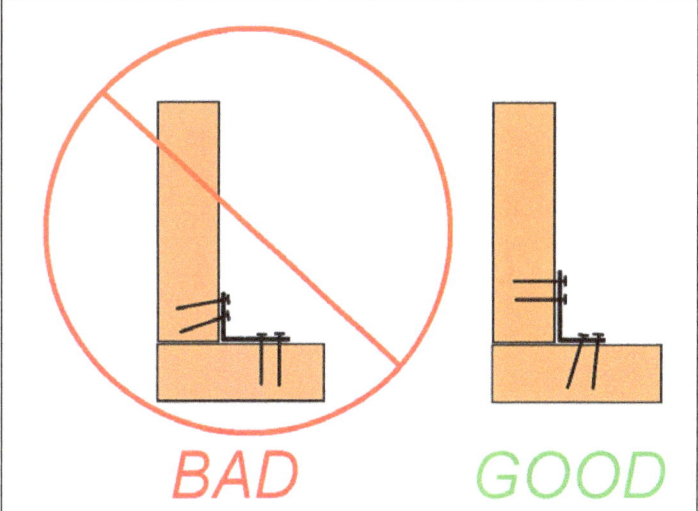

Figure 4-8 Install framing connector angles so you can drive the nails as far as possible from the edges of the wood. This type of connector is most often used to connect joist blocks or end joists to a mudsill or top plate; in such cases, place the long leg of the connector against the block or joist. This sketch does not show obstructions—nails usually have to be driven with a palm-nailer, which forces you to drive them in at an angle. CA50, L50, or AC5 connectors allow more room for nailing than the "A35" connector series.

4.3 *Practices & materials to avoid*

Some practices and materials should not be used in retrofit construction, and possibly construction anywhere. (In Section 8.1 we will look at bad installations of what may be good products or materials; this section addresses things that should almost never be used).

Do not use cheap plywood. Three-ply or four-ply plywood allows a greater chance that defects will align from one veneer layer to another. You will have no way to see these until the earthquake finds them.

Do not use oriented strand board (OSB). I used to like OSB, but my last personal experience with it was that it swelled up and began to delaminate after exposure to a single day of light rain—even though the panels were rated for weather exposure during construction. Home inspectors report frequent OSB failure.

Do not use lumber treated with ACQ, ACZA, CC, CA, or any other copper-based chemicals. Figure 4-9 shows the effect of commonly used pressure treated wood on a steel anchor bolt. This photo was taken about 6 years after construction. The white powder on the galvanized metal connector indicates that the coating is failing—what do you think the nails that secure the OSB to the mudsill look like? See Section 8.5

Figure 4-9 Recent chemical formulations containing copper compounds corrode steel fasteners very aggressively. USE ONLY BORATE-TREATED LUMBER where PT lumber is needed. See Section 8.5 Photo: John Fryer, GGASHI

Do not attach *shear panels* to framing members with adhesive. Current code prohibits using adhesives to attach shear panels to framing because it vastly increases the stiffness and reduces the ductility of the structure. **Note that the prohibition is against attaching *shear panels* to *framing*:** I do recommend adhesives in other connections, as specifically discussed in other sections.

Do not use drywall screws for any structural connections. Drywall screws are brittle and snap easily under load. Structural screws rated for loads are entirely different.

Do not use more nails than needed–especially for attaching shear panels. Using too many nails can split framing members, and unequal plywood edge nailing can cause early failure along edges with fewer nails.

One common misconception is that you need to nail plywood along each framing member behind the panel, such as built-up studs or wall top plates, at the specified edge nailing. The plywood does not know how the framing is configured behind it—all it knows is that it needs uniform nailing along each edge. Treat built-up members as if they are a single piece of lumber, and place ***one row*** of nails along each panel edge (you may stagger the nails to reduce the chance of splitting the wood).

Do not notch or slot shear plywood—or do any of the other things advised against in the *Shear Wall Guide*.

4.4 Know when to stop engineering and start building

Engineers are trained to analyze things. Sometimes we don't know when to stop. The following sections discuss a few cases where money is likely better spent on retrofit construction than retrofit design.

4.4.1 Retrofit work is rarely economical above the first floor

Sometimes I hear of engineers climbing into attics to investigate connections from the walls to the roof diaphragm. For masonry buildings the diaphragm connections are essential. In wood-framed houses, though, the chance of the original connections failing is essentially zero unless the relatively rare weakness discussed in Appendix F is present. [The National Research Council did extensive research after the 1994 Northridge earthquake. They noted "roof structural damage was almost nonexistent in single family dwellings," aside from chimney damage. Their report notes a failure of one roof where original light-weight roofing had been replaced with heavy tiles that overloaded the original construction. See Appendix G.]

Look at the photos in Figure 4-10 through Figure 4-12. These are spectacular examples of buildings that failed—but only *parts* of the buildings. The house in Figure 4-10 collapsed because of a weak-story garage. The row of windows in the living area creates a very weak condition at that level as well, but it did not fail—even when the house dropped a couple of feet. If you look carefully you notice that you can see all the way through a window on the far side of the house and into the forest beyond. There were very few cross-walls to stabilize the roof on this house; yet none of the windows even broke when it fell. Granted, the house may have landed softly on a parked car or two.

It can be argued that we don't really know how much shaking the main level of the house could withstand, because weak cripple walls absorbed all of the shaking. Look at the house in Figure 4-11; it dropped about 5 feet when the cripple walls collapsed, and yet none of the windows even broke. The house shown in Figure 4-12 rolled off of a house-mover's trailer and held together remarkably. Figure Figure 4-13 shows a house that fell off of its supports in the 1964 Alaska Earthquake. If you could right this house, it appears

Figure 4-10 Despite the extensive windows in this house, the upper level sustained little damage. FEMA photo by J. K. Nakata.

Figure 4-11 This house is held together after falling about five feet. FEMA photo.

Figure 4-12 Test of roof-to-wall connections, Canadian style. Photo by Thaddeus Holownia

largely undamaged. (This is an excellent example of where a few dollars in hardware to keep the posts in place, together with foundation anchors, could well have been enough protection.) To me, these are fairly good examples that original construction above the first floor is quite robust.

When you strengthen the connections between the lowest floor and the foundation, an earthquake's energy would simply find its way to the next weakest portion of the building. The analogy used is a chain being only as strong as its weakest link: strengthen the weak link and you still have to worry about the next-weakest one. Retrofit designers need to remember that the "link" represented by the cripple walls is **much weaker** than the rest of the building. Strengthening the next-weakest link often means making repairs in the living spaces; this is seldom economical.

Figure 4-13 This house fell several feet at one end, but the roof-to-wall connections (and others) did not fail. If the posts supporting the left side of the house had been secured it would likely have remained in good condition.

The question is not whether your first floor walls are "strong enough" to meet any sort of code; it's whether you can afford to strengthen them, and if the added strength is worth the price. Strengthening walls in the living area may be worthwhile in some cases if you have a weak wall line at that particular level of the house, but usually it is even more worthwhile to strengthen walls in the crawlspace below.

4.4.2 The "all-or-nothing" approach is rarely practical or affordable

Bringing a whole house up to the current code would be extremely expensive. This is especially true of older houses that were built before any codes were in place at all. Considering that a fully upgraded house may not comply with the next round of code revisions, what would you really gain from such a venture anyway? The greatest dollar amount of earthquake damage is caused by "moderate" events in the range of 6.0 to 7 on the earthquake scales; strengthening to resist this size event is much easier, and more effective over all, than trying to resist the "maximum considered earthquake," or MCE.

A design for a retrofit that would survive a mega-quake is useless if the owner can't afford to build it. Designing affordable steps to create a reasonable well-matched structural system is the approach I recommend. I give my clients a prioritized list of components that need strengthening; what they cannot afford to do might be a project that a future owner can accomplish.

4.5 Retrofit priorities

Retrofitting is expensive. Banks are starting to realize that lending you money for an earthquake retrofit to prevent a few hundred thousand dollars in damages is a good idea. Homeowners paying out of their own pockets may need to approach a retrofit one or two steps at a time. The following reflects general conditions I have encountered in the field. The reasoning behind the ranking is explained in more detail in the book sections noted. The recommendations apply only to houses on fairly level sites; see the restrictions noted below.

4.5.1 "Older" houses

This list applies to all older houses (pre-1960 or so), regardless of the shape of the house.

1. Remove any brick chimneys, especially if they were built with lime mortar (see Figure 4-14).
   There is no affordable way to effectively retrofit a masonry chimney. See Section 12.1.

   If you cannot remove your chimney at least down to the ceiling level, place ¾" plywood, or two layers of thinner plywood that total ¾" thickness on top of the ceiling framing around your chimney (keep any combustible material at least 2" away from the chimney). You will almost certainly have to cut the plywood into pieces smaller than 4'x8' in order to get it into your attic. Some experts recommend *not* nailing the plywood in place, but I disagree with this position. I would rather have the assurance the plywood will stay where I put it during a quake and not worry that I have made the ceiling diaphragm more rigid, when the original designers and builders gave absolutely no consideration to the ceiling's rigidity.

   You can keep the fireplace and transition from the masonry to a new metal chimney, or replace the fireplace with a manufactured unit. Figure 4-15 shows the remains of a chimney. Note the top portion of the chimney still connected to the roof, waiting to fall in an aftershock

2. If your house has cripple walls, install plywood bracing panels. If you cannot install plywood on each side of the house, install it on the narrowest sides first (assuming your house is basically a rectangle).

Figure 4-14 This is what happens to lime mortar after 100 years: it vanishes—even under paint.

Figure 4-15 Typical chimney damage. FEMA photo by Andrea Booher

3. Install shear transfer ties where joists are parallel to the supporting cripple wall or mudsill.
4. Anchor the mudsills to the footings. If you have cripple walls it is best to install anchors before the plywood bracing because you can use a wider assortment of anchors.
5. Install shear transfer ties from joists that are perpendicular to the supporting framing.

4.5.2 Houses built between 1960 and 1990

Since 1960 (at least in California) anchorage from wood framing to the foundation was fairly universal. As some problems were mostly eliminated, others were introduced. For houses built between 1960 and 1990, I recommend upgrades in the following order:

1. Remove masonry chimneys. The Northridge Earthquake tested thousands of chimneys built since 1950 (when the requirement for building chimneys with steel reinforcing bars was introduced). About 50,000 chimneys failed the test, even with the steel reinforcing, and Portland cement mortar (rather than lime mortar).
2. Add plywood bracing to cripple walls.
3. Install shear transfer ties to joists parallel to supporting framing.
4. For stucco houses, evaluate how the stucco lath was connected to the framing and retrofit if needed; see Section 11.5

4.5.3 Houses built since 1990

Since 1990 some new issues have arisen. For homes constructed to meet only the prescriptive codes between 1990 and 2025 (you read correctly: 2025) I recommend the following.

1. Evaluate the connectors used in pressure-treated mudsills. Much more corrosive chemical treatments were introduced in the mid 1990's. Beginning in 2004 the majority of pressure-treated lumber used for mudsills was treated with the much more aggressive formulations.

    Even code-compliant fasteners corrode severely in some—possibly most—cases. This topic is addressed in detail in Section 8.5. The connectors for many shear panels often did not even comply with the code requirement for hot-dip-galvanized material; in this case you may have no connections to the mudsill left at all.

    It may take decades for code requirements to catch up with this problem. In my professional opinion, anything besides stainless steel fasteners or connectors used in treated lumber constitutes a defective condition, except in the case of borate-treated lumber.

2. Remove masonry chimneys.
3. Install shear transfer ties to joists parallel to supporting framing.

    Code language does not yet require an adequate connection from end joists to framing below, or from joists to mudsills if there is no cripple wall. A proposed code change addressing this may go into effect as soon as the 2015 building code (which would take effect in 2017 in California).

4. Evaluate stucco as above

4.5.4 Exceptions

The preceding recommendations are intended for "regular" buildings, and could change substantially if your house has certain traits. You should seek the advice of an experienced engineer if your house has any of the following features:

- Soft or weak story created by living space above a garage
- Post-and-pier foundation system (where the house is supported *only* on isolated posts and piers, in other words does not have a perimeter foundation).
- Hillside construction, either on tall posts or steeply sloping or stepped footings
- Masonry veneer
- "Custom" architectural features
- Split-level construction
- Engineered designs; Since 1980, custom homes were designed by engineers with increasing frequency. Even the best engineered designs are subject to construction problems—that's why I wrote *Wood-framed Shear Wall Construction—an Illustrated Guide*. Unfortunately there are many newer houses that look good on paper, but the construction does not match.

4.5.5 Recommendations for all homeowners

- Make your house lighter; if your roofing is heavy tile, tar-and-gravel, or simply several layers of existing roofing, when it comes time to replace the roof use lightweight materials.
- Anchor tall, heavy furniture. About half of the earthquake deaths in the US are caused by falling furniture.
- Brace your water heater tank. Water heaters are hardly ever braced adequately. You should be able to push against the tank as hard as you can and it should not move.
- Install an automatic gas shut-off valve. Learn how to safely light pilot lights so you do not have to wait until the utility company lights them so you have heat and hot water.

Masonry fireplaces & chimneys provide the focal point of some houses. Fireplaces in general are expensive, messy, and inefficient to operate. Many areas in California restrict wood burning because it adds significant air pollution in the form of soot. If you would really miss your fireplace, consider converting it to a gas-burning insert with tight-fitting glass doors and a light-weight metal flue system.

I always recommend that people remove their dangerous, inefficient, polluting fireplaces. At least one of those attributes should prompt you to remove your fireplace.

Here's an account of what happened to one chimney during the 1989 Loma Prieta earthquake, from an experienced home inspector:

"My client who owns the house is a famous rock star. His father-in-law lives there, and is also a musician. When the shaking began, he was in his living room, sitting on the couch opposite the masonry fireplace playing his saxophone. He set the saxophone down on the couch and went toward the front door, and then the fireplace exploded. The mantle was ejected across the room and hit his saxophone. Pieces of the chimney were scattered over the roof, living room, and exterior."

Douglas Hansen, *Code Check*

- Install reinforcing film on large, untempered windows—especially windows near beds or cribs. One such film product is 3M's "Scotch-Shield."
- Have an emergency kit, food, water, and supplies to last two weeks. ABAG, the Red Cross, FEMA, and many local government agencies have online resources and lists of recommended supplies.

See Section 12.2 for more information on reducing "non-structural" hazards.

Chapter 5 Designing a Basic Earthquake Retrofit

Earthquake retrofit design includes analyzing the existing structure and then determining how and where to add strengthening measures. For experienced engineers, calculating the forces is usually easy compared to designing connections between various existing components and new structural elements or systems. However, explaining the analytical principles is more than I can, or probably should, address in this book: nobody should entrust their family's safety to their possible misunderstanding of fairly technical material.

All of the "standard retrofit plans" in use in the Bay Area, Los Angeles, Seattle, and perhaps other areas are intended for implementation by non-engineers. This section is intended only to explain the very basic concepts involved in calculating earthquake forces on a building.

The focus of this book is proper installation methods. Other resources have instructions and tables that will tell you how many mudsill anchors to install and how much plywood you need. One of these is *Standard Plan A* distributed by the Association of Bay Area Governments (ABAG). This document is usually referred to as "Plan Set A," and is available online here: http://www.abag.ca.gov/bayarea/eqmaps/fixit/Plan%20Set%20A.pdf. Seattle's *Project Impact*, and LA's *Standard Plan One* (http://ladbs.org/LADBSWeb/LADBS_Forms/Publications/anchor_bolting.pdf) also give information on anchor and plywood placement and quantities.

The standard plans listed above include four foundation connection details to wood framing (repeating these for both concrete and concrete block foundations gives them eight details). Early in writing this book I sat down with a pad and pencil and sketched almost 30 different framing conditions I have encountered; I have seen very few of these conditions addressed in other retrofit publications, and feel that this information is much more important to provide than repeating existing material.

If you can calculate the floor area of a house, you can use the tables in the prescriptive plans. However, implementing one of the standard plans is still best left to an experienced contractor. ABAG is currently updating an all-day training course intended to teach proper application of Plan Set A. There is a lot of information packed onto the two plan sheets, and a homeowner would need to be extremely dedicated to master all the material needed—let alone construct the retrofit.

5.1 Basic physical principles that govern earthquake forces

One of Newton's laws is *f=ma*, meaning *f*orce equals *m*ass times *a*cceleration. We can say that the earthquake force equals the weight (which is related to the mass) of the building times the acceleration, which is caused by the earthquake. If we know the house's weight and the earthquake acceleration we can easily calculate the earthquake force on the building.

Engineers determine the expected earthquake acceleration for a particular structure based on many factors. The most important are:

- How close the nearest earthquake fault is
- How strong an earthquake the nearest fault is expected to generate
- The soil or rock type and depth underlying the structure
- The structural system used to resist the earthquake forces

Other "safety factors" are applied to account for the importance of the structure and whether the structural system has redundant elements in case some elements fail, etc.

For wood-framed residential buildings braced with plywood panels, the earthquake forces usually fall in the range of 8% to 12% of the total building weight. This figure is lower than the standard plans use. There are several reasons for this; the most important one is the assumption that an experienced engineer will design the entire structural system for a building when the lower figures are used. For increased safety, non-engineered retrofit systems should use closer to 20% of the building weight as the expected earthquake force.

5.2 *Square buildings are easy*

When analyzing a wood-framed building, we consider earthquake forces acting in two principal directions: left-to-right, and front-to-back. If you have a square building with relatively uniform material weights throughout, and no interior bracing walls, it should be obvious that you will need the same earthquake resistance along each of the outside cripple walls.

As a quick example using the Reinforcement Schedule in Appendix A we can determine the plywood bracing that Plan Set A would require for a typical one-story stucco house with a 1500 square foot floor area. (This example is intended to show the difference between Plan Set A and other prescriptive requirements—don't worry too much about following along.) We go to the table in the Reinforcing Schedule (see Appendix A) and find the rows for 1500 square feet total floor area. Using the row for "heavy" construction (stucco exterior makes our house "heavy") the table shows we need 22'-8" of plywood bracing panels along each wall line. Other columns give the number of mudsill anchors and framing connectors.

5.3 *Rectangular houses are easy, too—maybe*

For a rectangular building we also assume that the earthquake force acting left-to-right would be the same as the force acting front-to-back. Therefore we need the same overall strength on the short walls as the long walls. This should also be obvious; despite this, some retrofit guidelines are not based on this reality. Depending on the overall shape and outline of your house, though, this contradiction will make sense.

Let's take another example house that has the same floor area as the one described above. The only difference is our new house is a rectangle, 20 feet by 75 feet. Plan Set A points us to the exact same length of plywood needed along each wall: 22'-8". But now we face the problem of installing 22 feet of plywood panels on end walls that are only 20 feet long.

Using the International Existing Building Code (IEBC) or variants such as found in Seattle or Los Angeles our retrofit would get even more strange. These codes would require that we brace 50 percent of the length of each wall line. Following this we would have 10 feet of bracing panels on the end walls and 37.5 feet of bracing on the long side walls. This makes no sense if you think only of the earthquake forces and strength of your bracing panels, but accommodates complex building shapes more easily than Plan Set A does.

5.4 Choose your code

The Association of Bay Area Governments developed Plan Set A to provide more sensible bracing for long, narrow houses commonly found in the Bay Area. Volunteers from the Structural Engineers Association of Northern California did a great deal of analysis to determine how much plywood and how many foundation anchors and framing connectors would be needed for houses of various sizes and weights. As demonstrated above, the plan set does not work for extremely narrow houses, where required plywood bracing lengths may exceed the actual length of the walls (this problem is more common for two-story houses).

Once you have a house with more than four walls, using Plan Set A can get very confusing; what is a "wall line," and how do you apportion the required total length of plywood bracing along it? Bracing a set percentage of wall length is easier to understand if you have even a moderately complex floor plan.

Code Recommendation:

- Use Plan Set A for simple, four-sided houses (square or rectangular). Even when the required length of plywood will not fit onto the end walls of a long, narrow rectangular house, Plan Set A makes a lot more sense than bracing only a percentage of the wall length.
- Use the IEBC or LA/Seattle standard plans for relatively square houses, or houses with complicated outlines.
- Hope for a new and improved guideline: As of this writing, FEMA is developing an "advisory" after the 2014 Napa, CA earthquake (tentatively titled "*South Napa Earthquake Recovery Advisory FEMA DR-4193-RA2*") The plan set associated with the advisory will attempt to improve upon and expand the retrofit methods presented in previous documents. The advisory includes retrofit methods for cripple walls up to 7 feet tall. Another goal is to address framing conditions that were common in older houses. A few of the retrofit connections in the FEMA document came from Chapter 7 of this book.

5.5 *"One size fits all" solutions often don't fit anything very well*

Any "prescriptive" retrofit code adopted in a democratic society will be some sort of mixture of ideas contributed by various interests—engineers, contractors, building officials, hardware manufacturers, etc. While each committee member may contribute useful ideas, those ideas do not apply to every situation. With a prescriptive code you can't always separate the ideas that don't make sense for your project.

The public expects any code developed by experts to yield relatively safe results when followed. Writing a code for new construction is easy compared to a retrofit code. How do you account for varying material quality, or possible hidden wood decay or rusting foundation anchors? —you add safety factors to your design. By the time each committee member adds his or her own safety factor, each of which is perfectly reasonable by itself, you may end up with a retrofit that is overly conservative.

Plan Set A is an example of multiple layers of pessimistic assumptions. You could build a new house to meet the current code requirements that would have less plywood than Plan Set A requires for a retrofit.

I recently compared the length of plywood bracing needed for a retrofit following Plan Set A with the bracing needed to meet the minimum IEBC requirements following sound engineering practices. The Plan Set A retrofit would have required twice as much plywood as my engineered solution and 24 additional foundation anchors.

For those who can afford only the minimum level of protection the IEBC suggests, I have concluded that the most cost-effective retrofit in most cases is a design engineered to meet the IEBC requirements. This does not mean you get equal *protection* for less money; it means you get *some* protection for less money. If you need to meet some minimal eligibility criteria to qualify for an insurance rate reduction (as may apply in California in the near future) or a tax rebate (currently available in the City of Berkeley when a home is sold) or the like, an engineered retrofit design may be the most economical way to meet those minimum criteria.

I should emphasize that a Plan Set A retrofit is more expensive primarily because it requires more plywood and foundation anchors than an engineered system would. The additional plywood and anchors would usually make the Plan Set A retrofit stronger than the minimum needed to meet the IEBC. They may also make the retrofit unaffordable.

5.6 *"Plan Set B"—the inside story from a committee survivor*

In the spring of 2015 I worked as a subcontractor on a FEMA project that expands Plan Set A. The project team and peer reviewers were all engineers at the top of the profession, with great hopes to produce a document that would apply to a majority of older homes. We quickly realized that producing a single "standard plan" that would apply to houses ranging in age from 30 years old to 130 years old, of varying architectural and construction styles, was not a reasonably achievable goal. Here's a partial list of obstacles we faced:

- Foundation materials—could we include brick or stone masonry, or concrete block? Retrofit anchor manufacturers don't publish load values for anything except relatively new concrete or reinforced concrete block. How could we be sure that concrete block would be reinforced adequately? We had to rule out everything except concrete.
- Footing size and shape—what size footing could we count on for anchoring a tie-down? Contractors reported that total footing height (from bottom of footing to bottom of mudsill) is often only 14 inches in the class of houses we were most concerned about. That meant we had to lengthen the wall bracing lengths to limit uplift at the ends of the walls.
- Material weights—plaster on walls is generally thought to be ½" thick. Sometimes it is, but most of the time I see ¼" to 3/8" thick wall and ceiling plaster. This would substantially reduce the calculated earthquake forces. Accounting for this would require yet another table showing how much plywood and how many mudsill anchors to install.
- Shape of the house—engineers usually follow the "tributary area" method for determining how much force each particular braced wall needs to resist. Without explaining this and a few other topics, any house with more than six corners will challenge most users of the plan set.
- Lumber size—sometime around 1930, rough-cut lumber was crowded out of the market by dressed lumber. Rough-cut lumber is usually 1-3/4" to a full 2" thick, and requires longer connectors when fastening through it. Dressed lumber is 1-1/2" to 1-5/8" thick, and accepts the nails easily found today. Now we need to include 20-penny nails among the fasteners, or make other accommodations for thicker old lumber.
- Multiple framing methods—Plan Set A included about seven common connection details. Working only in a limited part of California I have encountered over 30 connection variations, some of them much more complex than the ones addressed in Plan Set A. Some of the details I developed to accommodate these less-common conditions were adopted into the FEMA plan set, but the committee did not have time to carefully evaluate and approve many of the details I offered. Including too many details also becomes confusing for users of the plan set.
- Different stock hardware items—we did not want to provide generic descriptions of hardware and leave the reader wondering where to get a connector rated for a particular load. This meant comparing hardware from at least three manufacturers (to comply with federal regulations avoiding favoritism). We found that the rated loads were usually different for similar connectors; this meant we had to choose the lowest load rating to use in our calculations, or provide connector quantities for each brand of connector.

A "standard plan" that would cover even the partial list of conditions above would need to be hundreds of pages long, and you would need to pick out which pages were applicable to your project. The other alternative is to create dozens of standard plans: for homes with older concrete footings, reinforced concrete footings, hillside homes, balloon-framed construction, etc., and then multiply those to account for several different ranges of building material weights, number of stories, etc.

Chapter 6 <u>Detailed Construction Methods</u>

Detailed training materials for contractors and builders have been outpaced by engineering analysis methods in the last 20 years or more. The main purpose of this book is to present effective earthquake retrofit *construction* methods. The following two chapters give specific information and examples of these methods. The methods shown are *not* the only ones that will work. Other engineers will come up with various ways of resisting earthquake forces—it's what we do.

Chapter 7 includes construction details for many of the methods shown in this chapter.

6.1 <u>Jobsite safety</u>

Construction work is dangerous. Construction work in cramped, unventilated areas, in awkward and uncomfortable positions is especially dangerous. Anyone undertaking construction work needs to accept responsibility for his or her own safety. Proper safety gear is an urgent necessity. Describing such gear is beyond the scope of this book.

6.1.1 <u>Things to fear on any construction site</u>

- **Power tools:** Learn about the tools you are using. Tie back long hair so it doesn't get caught in rotating machinery. Be alert and careful.
- **Noise:** Wear earplugs, over-the-ear hearing protectors, or preferably both. Palm nailers are some of the loudest tools made (I have operated pneumatic jackhammers and thought they were loud until I heard a palm nailer).
- **Dust:** Wear a respirator.
- **Sharp things:** Wear gloves.
- **Flying debris:** Wear safety glasses (I think you get the idea…)

> *Fear itself* Some crawlspaces are more accommodating than others. I've worked many days under houses. For one evaluation, I slithered into an 18" crawlspace, back past two interior cripple walls (not post-and-girder interior supports, but studs on 16" centers, which do not provide enough space for an adult to fit between them: like prison bars.....) To reach the very far reaches of the crawlspace, I had to put one arm out in front of me to fit through the third access opening on my journey, each of which was smaller than required by code. The opening was just beyond a metal furnace duct that was a little less high off the ground than I am thick. Prior explorers had rubbed the insulation off of the duct and creased the metal. When I was half way through the access hole my jumpsuit snagged on something, and I began to wonder how I would die if the Hayward Fault broke loose at that moment…. suddenly, by having the house fall on me? or slowly because a gas line broke and I suffocated? or the house caught fire…. Or would I just die slowly from dehydration? I didn't panic, but I was glad to finish my exploration and head back toward daylight—and freedom.
>
> I told an experienced retrofit contractor about the episode. He laughed and said, "Try installing a UFP10 while you're in that position!" There are times when you should call someone like him to do the crawling—it's worth it.

6.1.2 Things to be afraid of under old houses

- **Pesticides and other chemicals**—The stuff sprayed under houses to kill termites, ants, spiders, etc. used to be even more deadly than the chemicals allowed today. Sawdust from treated wood is hazardous.
- **Critters**--Rats, squirrels, possums, cats, snakes and other creatures live under houses, and die there, and poop there. I have crawled around a few pretty disgusting things, and probably crawled over lots of stuff I'm glad not to know about.
- **Hanta virus**—carried by mice; can be deadly.
- **Parasitic worms**—carried by raccoons and possibly other hosts. Wear gloves and wash your hands so you don't ingest the eggs.
- **Electrical hazards, sharp things, sewage leaks**—Home inspectors find live, uninsulated electrical wires dangling in crawlspaces more often than you would hope. Sometimes the wires are in contact with moist ground, which can create an electrocution hazard beyond the wires themselves. Rusty nails, broken glass, splinters, sharp pieces of metal, etc. are commonplace under houses.

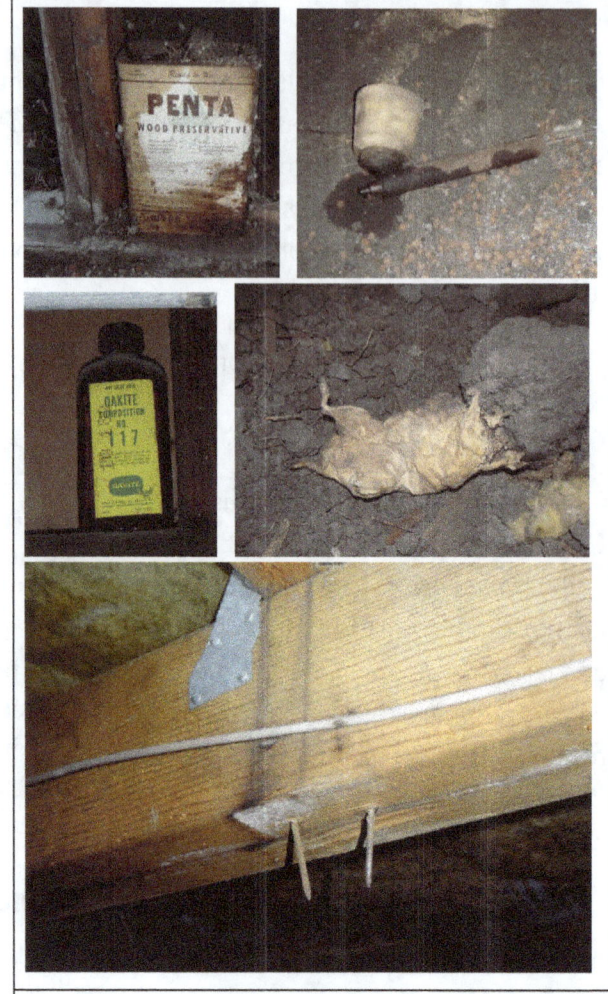

Figure 6-1 A sampling of nasty things you find under houses. Photos: Bottom—author; others: Paul Barraza, GGASHI

6.2 "Standard" Construction Conditions

This section addresses framing methods that were commonly used in houses since the 1870's or so. Practices varied regionally, and some builders were no doubt slow to move away from timber-framing methods. Your house may be special.

6.2.1 Connecting Existing Mudsills & Foundations

Mudsills provide the "reaction" force for the retrofit shear wall; they must be connected to the footings in almost any retrofit, or other connections created. In the SF Bay Area mudsills are almost universally full-dimension, clear, close-grain redwood. The fine grain of the wood accepts nails very well without splitting. The full 2-inch thickness (sometimes more) provides additional holding power for retrofit mudsill anchors. Redwood is not toxic like new preservative-treated wood can be. If possible I always try to use the original mudsills in a retrofit design.

The following sections address different connector placement, various types, how and where to install them, and the advantages of each.

6.2.1.1 Anchor location in mudsill

Bolts, screws, nails, and other fasteners must all be installed so they do not break out of the members they connect. You have to consider fastener placement in both the mudsill and the concrete.

Placement requirements recently changed for bolts connecting wood members. The current codes require installing bolts at least 12 bolt diameters from the end of a mudsill (increased from 7 diameters in past code editions). Too many mudsills had bolts break out at the ends during recent earthquakes. Anchors are also required within 12 inches of the end of a mudsill. For ½" diameter anchors this means you need to locate anchors between 6 and 12 inches of each end; for 5/8" anchors you need one between 7½" and 12 inches of the ends. See Figure 6-2 and Figure 6-3. The retrofit codes (IEBC and offshoots) require anchor end-distance to be between 9" and 12", which makes code-compliant anchor placement impossible if a cripple stud falls in that range. Oops.

Figure 6-2 Retrofit codes require mudsill anchors to be placed between 9" and 12" from the ends of mudsills. The code for new construction requires an "end distance" of at least 12 bolt diameters, which allows more leeway for placing the anchors.

Figure 6-3 This installation does not meet the code. The anchor on the left is slightly too far from that end of the mudsill; the anchor at the right is too close to the other end.

Some retrofit installations include blocks nailed or screwed to the mudsill. Anchors are usually installed through the blocks. In this case, the anchor end distance in the blocks is not important (end distance is still important in the blocks, but it is based on the fasteners from the block to the mudsill; these are typically much smaller in diameter and require an end distance of two or three inches). Figure Figure 6-4 shows an anchor bolt being installed far from the ends of the mudsill.

You also need to maintain certain clearances to the edges of the mudsill. For loads acting along the length of a member, as occurs in a mudsill along a shear wall, bolts must be installed at least 1½ diameters from the member edges. This works out to ¾" and 15/16", respectively, for ½" and 5/8" bolts—however, since we need to install 3" square plate washers on top of mudsills, the practical minimum edge distance is 1½."

Bolt spacing along the length of the mudsill must be at least 4 bolt diameters. There is generally plenty of room to meet this requirement. If we use plate washers, that forces us to have at least 3" between bolts.

We also need to consider edge distances to the concrete; this varies depending on the type of retrofit anchor, as discussed in the following sections. The anchor manufacturer's instructions take precedence over my general recommendations.

Figure 6-4 Foundation anchors connect the MUDSILL to the foundation, not the blocks. The anchor end-distance that matters here is in the mudsill. This previously-retrofitted cripple wall is receiving additional anchors, and an existing anchor obstructed placing the new anchor in the center of the block (which would be preferred when possible).

The preceding focuses on one connector at a time—where can we install anchors along the shear wall? Shear wall schedules often included in plans may give bolt spacing requirements for various different shear plywood and nailing patterns. Almost universally, the bolt spacing applies to only the length of mudsill that has plywood installed, and *not* to the entire length of mudsill. For example, if the plans show an 8-foot long shear wall that requires bolts every two feet, you need to install bolts at two-foot spacing along only the 8 feet of shear wall even if the mudsill is, say, 24 feet long.

The preceding does not mean we can install anchors *only* where the shear plywood goes. A mudsill can act as a collector (See Chapter 4 of the *Shear Wall Guide* for an explanation of collectors) and distribute loads along its whole length—whether or not plywood is attached at any particular region. There are several good reasons to distribute the anchors along the mudsill:

- It is easier to verify anchors that are not hidden behind plywood (even if inspection holes are provided)

- Spreading anchors out along a greater length of foundation, especially brick foundations, distributes loads more evenly
- You have a better chance of avoiding obstructions
- Some anchors (such as the UFP10 and RFA series) obstruct installing the shear panels
The number of anchors you can install outside the length of the shear wall depends on the size and strength of the mudsill; for full-dimension 2x4 redwood, probably no more than three anchors should be placed outside of the shear wall at either end. The principles used to arrive at this are the same ones used to design collectors, which are discussed at length in the *Shear Wall Guide*.

I find it much more reliable to indicate the required number of anchors than call out a particular spacing; for retrofit work there are usually enough obstructions that hoping for uniform spacing is not reasonable.

6.2.1.2 Anchor location in concrete

Edge distances for different sorts of concrete anchors vary from about 1-1/2" to 4" or more. This can limit your choice of anchor types; expansion anchors typically need the greatest edge distances. Edge distances are discussed further in upcoming sections.

Various anchor types also have different requirements for embedment depth into the concrete.

6.2.1.3 What is the minimum concrete thickness needed for anchorage?

Sometimes you have a partition wall in a basement that looks like a candidate for a shear wall—until you start wondering if there is a footing under the wall.

For decades the building code has required 7-inch minimum embedment in concrete for mudsill anchors. (Note: this is for anchors cast in new footings. Retrofit anchors are specifically tested at shallower embedment depths, and achieve rated loads at whatever depth the manufacturer lists in their installation instructions.) Usually we are anchoring a mudsill to an existing footing and have no trouble getting the required embedment.

Sometimes the only reasonable existing wall is a non-bearing partition wall that was installed 50 years ago on top of a floor slab. Will this work for resisting shear? The short answer is, "yes, it will resist *some* shear." How thick is the slab? How close together are the cracks? How big is it? Is it surrounded by footings that will keep it from sliding? Do you have some means to resist the braced wall's tendency to overturn?

Using an existing floor slab for anchorage is better than nothing—but not as good as what I would like to see in a complete retrofit. Count it as a good first step; then come back when you can afford to and cut out the existing concrete, dig out for a real footing and place concrete as needed.

6.2.1.4 Retrofit "Bolts"

Besides standard mudsill anchors used in new construction there are four relatively common sorts of retrofit anchors (technically none of these are *bolts*, but we call them that anyway).

Since 1994 the building codes have required mudsill anchors in high seismic areas to be 5/8" diameter. This should not apply to buildings constructed decades earlier than that, but some jurisdictions insist on it. Anchors must not be installed too close to the edge of either the mudsill or the footing, or they may tear or break out under load. Larger diameter anchors require greater edge distance. Requiring 5/8" diameter anchors reduces the cases where you can use a retrofit bolt at all, thus making the designer's and installer's jobs harder.

Usually retrofit anchors are installed through an existing mudsill. Once you establish the anchor locations you need to drill through the wood with a wood-boring bit that will allow clearance for a concrete drill of the required diameter for your particular retrofit anchor. DO NOT drill through the wood with the same tool that you use for drilling into the concrete; doing so would mash the wood fibers and can greatly reduce the capacity of the anchor.

For new construction, codes require bolt holes bored in wood members to be 1/16" larger than the bolt diameter. Depending on the type of anchor you are installing, this may be easy to accomplish. For mudsills in new construction the bolt holes are almost always oversized to make it easier to drop the mudsill over all the bolts cast into the footing. In the case of retrofitting a mudsill that is already sitting on the footing it is easy to install a retrofit anchor in the code-prescribed hole size, or even a hole that is the same size as the anchor diameter. Requiring the 1/16" oversized hole does not make much sense in this case—especially if one considers that when adhesive anchors are used it is likely that the adhesive will fill up the small gap between the wood and the anchor.

Beginning with the most common types, the various retrofit anchors are:

Expansion anchors: Often called "red-heads," after the most widely known brand. Several models of expansion anchors are shown in Figure 6-5. Expansion anchors are the least expensive in most cases. They work fine when loaded in shear, and are appropriate for securing mudsills to footings in cases where required edge distance is available.

By their very principle, expansion anchors can break the concrete. Expansion anchors may have to be installed further from the edge of existing concrete than other anchors; these requirements are given in the manufacturer's information.

Expansion anchors should be avoided in cases where they are subject to direct pull-out of from the concrete, as is typical for tie-downs. If the existing concrete is of lower quality, adhesive anchors are more desirable in any case.

Expansion anchors usually are installed into the concrete in holes that are the same diameter as the anchor itself. This means that the bolt hole you drill through the mudsill will be 1/16" larger than the hole in the concrete.

Anchors are pounded into the hole with a hand-sledge. Since this process can deform the threads at the top of the anchor it is usually advisable to install the toothed washer (if used—see Section 6.2.1.7), plate washer and nut before driving the bolt into place. Once set to the proper depth, tighten the nut to the torque required by the manufacturer to set the expanding sleeve or wedge tightly against the inside of the hole in the concrete.

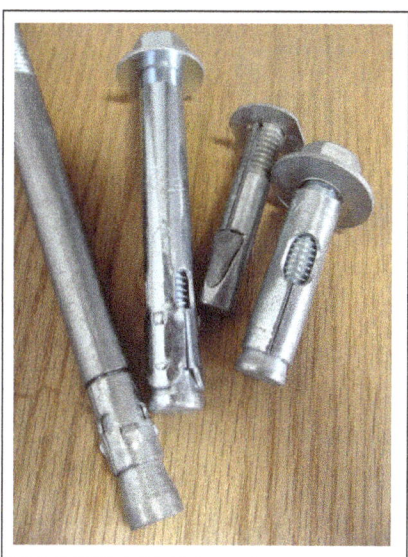

Figure 6-5 Expansion anchors of several types. For connecting mudsills, only the anchor on the left is long enough to embed in the concrete; the other three may be appropriate for connecting steel hardware (such as UFP10s or RFAs) to footings.

Adhesive anchors: These include epoxy, vinyl-ester, and other adhesives used to anchor a length of threaded rod. Adhesives are not all rated for pull-out when anchoring to cracked concrete. If you are using an adhesive anchor for a tie-down, make sure it is rated for tension forces in cracked concrete. Since adhesive anchors do not work by expanding against the sides of the hole, they are more gentle on old concrete than expansion anchors. Figure 6-6 shows sets of threaded rods with nuts and washers ready to be paired with adhesive.

Adhesive anchors are messy to install. The adhesive may take a long time to cure (overnight in some cases) before nuts can be tightened. Some adhesives smell really bad and will irritate your skin. On the plus side, the adhesive can fill over-sized anchor holes drilled in mudsills.

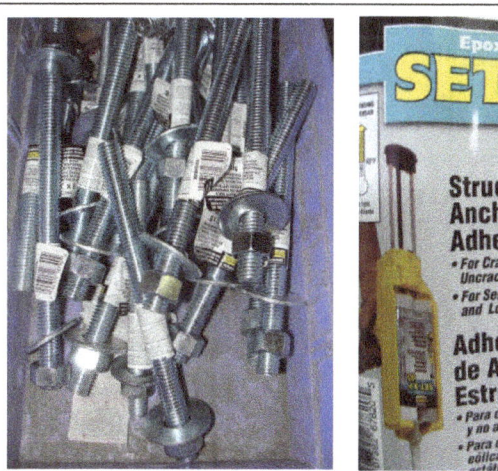

Figure 6-6 Adhesive anchors; preferred for masonry or weaker concrete. Edge-distance requirements are usually less than required for expansion anchors. In 2014, cost of the above adhesive alone was around five dollars per anchor.

Manufacturers have different requirements for installation depth and hole diameter for the various adhesive types. Some anchors can be used in holes that are 1/16" larger in diameter than the anchor is; in this case, the hole you drill in the concrete is the same diameter as

61

the hole through the mudsill. In some cases the hole in the concrete must be 1/8" (or more) larger in diameter than the anchor. For proper mudsill connection you will need to fill the space between the anchor and the wood with adhesive—which is easy enough to do by using plenty of adhesive when you install the anchor.

Before injecting or pouring adhesive into the hole it must be cleaned to full depth with a round, nylon bristle brush, and the loose dust blown out with compressed air. This is a very important step, and can complicate installation in cases where you must drag an air-hose into far-away reaches of the crawlspace. A contractor reported failing an inspection because he did not have his cleaning brush on site when the inspector came, to prove that he had properly cleaned the holes—never mind that the holes were, indeed, clean—as demonstrated by rubbing the threaded rod against the walls of the holes and bringing it out with no dust in the threads. Lesson: keep your brush on site until you pass inspection. (Note: Hilti Corporation recently introduced an adhesive listed for use *without* cleaning the holes first—but you have to use a special, expensive drill and vacuum system.)

A possible sub-category of adhesive anchors would be grouted anchors. Non-shrink grouts such as "Pour-Stone" (Custom Building Products) and "Rockite" (Hartline Products) mix with water and set to very high strength in less than an hour. It is hard to find load ratings for these products, which may make some people or building inspectors hesitant to use them. For loading in shear only, I believe grouted anchors are entirely suitable. They are commonly used to anchor machinery that produces millions of cycles of vibratory loading.

Threaded anchors: Figure 6-7 shows typical threaded anchors. This type of anchor came on the scene in the 1990's. They work well in shear or withdrawal, and you don't have to wait for smelly, messy adhesives to cure. The anchors themselves are expensive, but some contractors use them exclusively because they install quickly and simply; you do not need to wait for adhesive to cure.

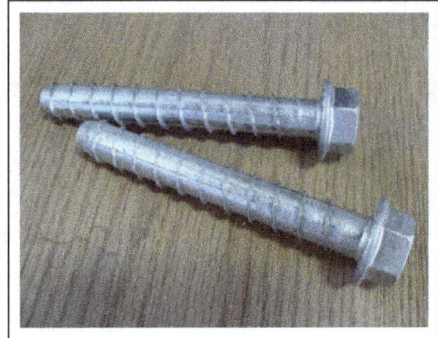

Figure 6-7 Threaded concrete anchors. These also need less edge distance than expansion anchors. Anchor cost is high, but installation is quick with an impact-driver; you do not need to wait for adhesive to set up.

Threaded anchors are typically installed in holes of the same diameter as the anchor itself. Holes must be drilled at least one anchor diameter deeper than the length of the anchor to allow for cuttings from the hardened steel threads to accumulate without blocking insertion of the anchor.

Under-cut anchors: These are becoming more common in response to the problem of other anchors (particularly expansion anchors) pulling out of concrete that cracks at the location of the anchor. I have not seen these anchors installed "in the wild" yet, but according to manufacturer's testing they appear suitable for most retrofit needs.

Some under-cut anchors require special drill bits to drill the under-cut hole; others have cutters built into the anchor (adding cost to individual anchors, but saving money in some cases).

Manufacturers offer different anchors depending on whether the component that needs anchoring is in place already, or will be installed after you set the anchor. For mudsills you will typically need the anchors made for the former condition; for other hardware the "post-installed" variety may not work well.

6.2.1.5 Anchor installation in restricted spaces

Installing and anchoring mudsills is easy until you build a house on top of them. After that, trying to add more anchors can become quite difficult. If you are adding retrofit bolts, you need to drill a vertical (or almost vertical) hole through the mudsill and into the concrete. Drilling equipment needs about 24" of overhead clearance. Sometimes you can cheat a little and work with the drill body up in the joist space, leaning the drill against the cripple wall top plate. If the cripple walls are shorter than 22 inches, though, drilling for retrofit bolts is not an option.

Leaning retrofit bolts more than a few degrees reduces their capacity significantly. The bolt in Figure 6-8 is useless for resisting loads. If you need to tilt an anchor, use a beveled washer under the nut to provide solid bearing, as shown in Figure 6-9 Beveled washers are commonly used to connect structural steel channels and I-beams; steel fabrication shops should have them. Ashby Lumber in Berkeley carries them; other good building supply centers probably do as well.

Figure 6-8 BAD: Leaning anchors lose their effectiveness. A "UFP10" would be better in this situation. Photo: Paul Rude, GGASHI

Figure 6-9 A beveled washer compensates for bolts that lean up to about 10 degrees.

6.2.1.6 Plate washers (also known as "bearing plates")

Many mudsills split lengthwise during recent earthquakes. The two leading causes of these failures are plywood uplift along the mudsill edge, and "punching" by the washers.

When an earthquake loads a shear wall, the whole assembly of plywood panels tends to rotate slightly like a giant square wheel. The rotating plywood panels lift up the mudsill at one end of the shear wall (and then the other, when the earthquake cycles back the other direction). Testing showed that standard round washers allowed the mudsill to lift up off the footing, which split the mudsill lengthwise by "cross-grain bending" (for further information see Figure 10 in the *Shear Wall Guide*). Plate washers installed along shear walls tend to reduce this problem; they work best when installed within ½" of the plywood, which is required by code.

Figure 6-10 Plate washers help keep mudsills from splitting lengthwise. (Note: This installation does not meet the code requirement for placing the edge of the washer within ½" of the face of the member to which the shear panels will be nailed.)

The code requirement that plate washers be installed within ½" of the edge of the mudsill to which you will attach the plywood is relatively unknown. Figure 6-10 shows the mudsill in a public library under construction where this requirement was missed by the designer, builders, City inspectors, and probably some independent inspection agencies. (In this case I would worry more about the shear panel edge nailing driven into the copper-containing treated mudsill—see Section 8.5.)

Note that the tests where round washers failed were mostly done on 8-foot tall walls. In shorter walls the rotation—and therefore mudsill uplift—would be less.

The other way the mudsills split was under extreme loading that bent the anchor bolts over. As a bolt bent, the nut and washer securing it rotated along and crushed into the top of the mudsill. In extreme cases, this punching helped cause mudsill splitting. This should not be a problem for mudsills secured with sufficient numbers of anchors.

Are plate washers really that important? Maybe not. One study concluded that shear walls *"which were identical except for washer size showed similar performance based on peak capacity, deflection at peak capacity, maximum deflection, and observations of failures at the conclusion of testing."* "Cyclic Tests of Engineered Shear Walls with Different Bottom Plate and Anchor Bolt Washer Sizes (Phase II)" by Rakesh Gupta, Heather Redler, and Milo Clauson, Oregon State University, 2007
www.awc.org/pdf/OSUFullScaleShearwallTestReportII-07.pdf

Use the hot-dipped galvanized washers; as of September, 2012, a contractor reported that they cost three cents apiece more than plain steel when purchased in bulk.

In many previously-retrofitted cripple walls, standard round washers are the only "defect." (Note: this same 'defect' is present in any house with shear walls built before 2002 or so....) For walls that are at least twice as wide as they are tall, I do not feel it is worth the effort to tear off existing plywood just to add plate washers on otherwise adequate foundation anchors. Patching plywood is problematic, and tearing it all off and replacing it damages the existing cripple studs.

6.2.1.7 Toothed washers increase connection capacity

Figure 6-11 shows a "mud-sill plate" (MSP) washer. The prongs at each corner embed in the grain of the mudsill and increase the holding power of the connection. The manufacturer who developed this connection had them tested and found them to increase the capacity of a bolt by 57% when used between the footing and the mudsill (prongs pointing up) and by about 25% when the MSP was placed on top of the mudsill with the

> The first requirement for plate washers appeared in the 1997 UBC, for 2-inch square, 3/16" thick washers. The current requirement for 3" square by 0.229" thick washers appeared in the 2007 California Building Code. (Standard round washers are still allowed in low seismic risk areas, which account for very limited areas of California.) Some building departments had interim requirements for 2½" square washers.
>
> Note that the minimum thickness is very close to ¼", being only $1/48^{th}$ of an inch thinner. (0.229" corresponds to 3-gauge steel plate; and people think that the metric system would confuse them?) Requiring ¼" thick washers will mean that none of the mass-produced washers will be technically correct.

prongs down. (When used on top of the mudsill, a plate washer must be placed over the MSP—it does not take the place of a plate washer.) When used both top and bottom of the mudsill, the MSPs almost double the shear capacity of a bolted connection. Obviously in retrofit situations you cannot install MSPs under the existing mudsill.

The original manufacturer of this great product went out of business. Currently KC Metals is the only manufacturer that makes the MSP, and possibly only if you order in very large quantities—like 50,000 at a time. Call your favorite hardware supplier and ask about these.

One consequence of this orphaned product is that the code approvals based on test results mentioned above have expired. This means that building departments have no current load ratings for the MSP. This should not be a problem if you are installing a "voluntary strengthening" system—but if you need to justify calculations based on using MSP's you won't get far. Until then, for $25 in materials you can increase your house's mudsill anchorage capacity by 40 percent, even if the building department won't believe it.

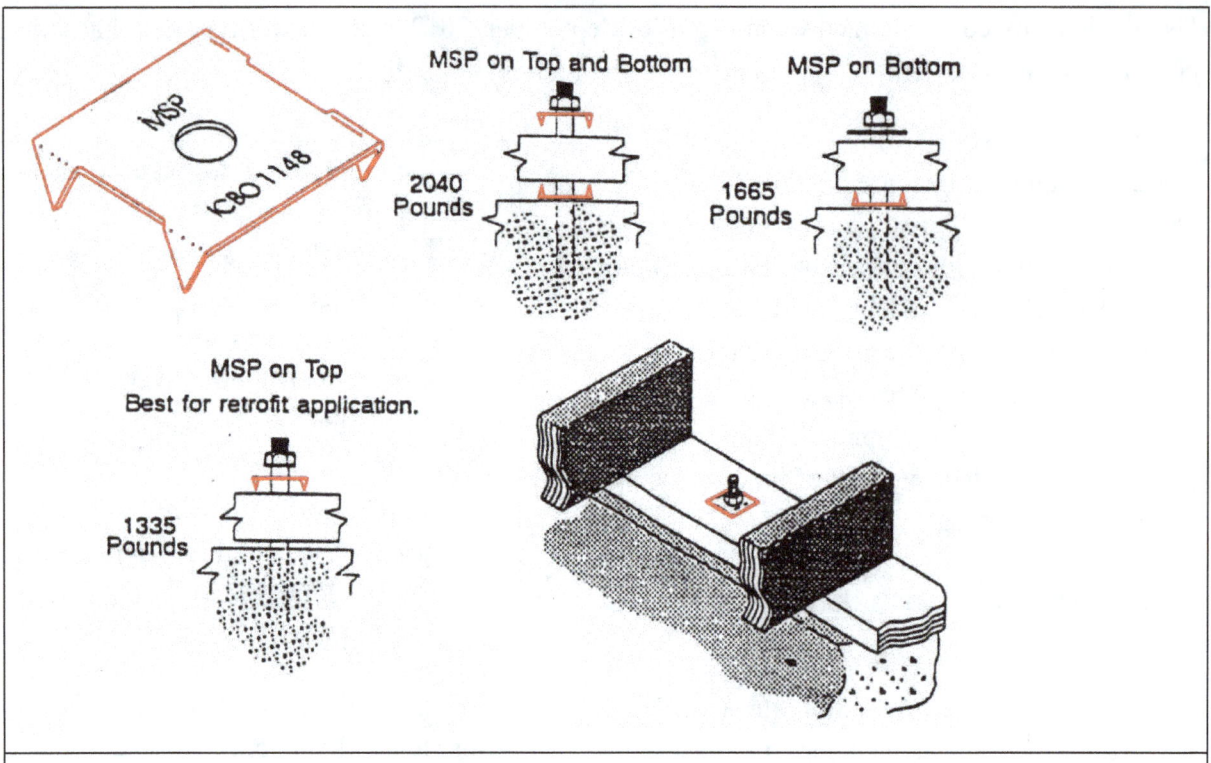

Figure 6-11 "Mudsill Plate" (MSP) usage and benefits

6.2.1.8 Use of existing mudsill anchors

Anchorage requirements for mudsills increased gradually over the last 70 years. Houses constructed since 1940 or so often have some anchor bolts that are in good condition and could—possibly—contribute earthquake resistance. Plate washers (and MSPs if available) should be installed on existing bolts; existing anchors should be ignored if the nuts cannot be removed because of excessive rust. If nuts on the existing anchors can be removed then the anchor is probably in good condition. However it is advisable to test the integrity of anchors by tightening them using a wrench with a 12-inch long handle and tightening the nut with all your strength. If the bolt does not twist off it is likely strong enough to carry earthquake loads as expected.

> MSPs do not have a current "listing report" that building departments require before they will accept a proprietary product. All the MSPs that are already installed did not vanish in a puff of black smoke when their approval report expired, though; it's just that the current manufacturer does not want to spend $10,000 to get the listing report renewed.

During the 1940s another way that mudsills were anchored was with ½" diameter rebar extended out of the footings. Mudsills were drilled to fit over the rebar, and then the bars bent over to hold the mudsill in place. Sometimes standard anchor bolts were simply bent over instead of having nuts and washers installed; see Figure 6-12. You do not want to rely on such anchors to provide much earthquake resistance. I recommend ignoring such anchors.

In addition to the above, existing anchors should not be considered toward the retrofit requirements if any of the following apply:

- Anchor is too rusted to allow installation of new nut and plate washer (a sign that the portion of the anchor hidden in the mudsill has probably also rusted)
- Anchor is set too high so nut cannot be tightened down on washer. (Shimming under the nut with no more than two plate-washers is probably acceptable.)
- Severely rusted anchors (shape of anchor threads or nut has begun to change). In this case, water-intrusion problems should be investigated.
- A gap of more than ¼" between the concrete footing and the mudsill at the anchor location.
- Concrete around anchor has spalled, cracked, or deteriorated to the point that anchor is visible instead of being fully embedded in concrete.
- Anchor is not solidly embedded in footing (wobbles or twists in concrete)
- Anchor is tilted more than 1 inch in 6 inches (approximately 10 degrees)
- Centerline of anchor is less than 6" from end of mudsill, less than 1" from edge of mudsill, or less than 1½" from face of concrete
- Existing expansion-type anchor that is loose in concrete and cannot be tightened
- Other conditions as shown in Figure 6-13

Figure 6-12 These anchorage methods were used by rebellious builders for a time.
Top: anchor bolt bent over mudsill (often the bolt was not even 'secured' with nails). Photo by Max Curtis, GGASHI.
Bottom: The mudsill in this photo was placed over a steel reinforcing bar that extended from the foundation. The bar was then bent down onto the mudsill.
Photo By Paul Rude, GGASHI.

The first time "foundation bolts" were required in the code, there was no mention of nuts and washers. Some ornery contractors refused to install nuts and washers on the bolts, claiming that the connections shown in Figure 6-12 met the letter of the code. Thousands of homeowners need to deal with what these contractors smugly got away with 60 or 70 years ago.
(*Thanks, guys; I hope it was really satisfying doing lousy work just to prove a point.*)

Anchor was set too low in the footing and the mudsill has been notched more than ¼" deep for installation of nut & washer Photo by John McComas	Anchor diameter is less than ½"	Oversized hole in mudsill (more than 1/16-inch greater than anchor diameter) unless space between anchor and wood is filled with epoxy or grout
	Foundation anchors were often poked into wet concrete. This commonly left a gap between the anchor and the concrete. The photo at left shows such an installation after removing the mudsill—the anchorage point to the concrete was effectively about an inch below the top of the stemwall. The extended anchor length would bend very easily under sideways loading. (Note: Even though engineers don't rely on friction, the rough concrete surface would provide substantial grip if the mudsill was clamped against it. This is not guaranteed, though.)	
		Arrow indicates anchor exposed between top of concrete and bottom of mudsill
Mudsill was notched to slip around anchor (mudsill not installed down over anchors—common when sill replaced during pest damage repairs). Photo by Paul Rude	Sometimes wet concrete settles below the mudsill level. This also leaves an extension of the anchor that will bend more easily than a direct, solid connection from wood to concrete.	

Figure 6-13   Examples of why existing anchors are often unreliable.

6.2.1.9 Foundation plates: UFP10s, RFAs, etc.

Beginning around 1940, many houses were built over low crawlspaces with no cripple walls; the floor framing is supported directly on the mudsill. This prevents you from installing additional mudsill anchors from above. As described earlier, cripple walls less than a couple of feet tall also prevent easy access for drilling for foundation anchors. Hardware manufacturers addressed this problem with several different products that attach to the inside face of the concrete foundation and to the top or inside edge of the mudsill, as shown in Figure 6-14.

Several types of connectors are available that attach to the side of the foundation with bolts into the concrete and structural screws into the edge of the mudsill. This hardware is specifically engineered for this situation, and has undergone testing to establish its load capacity. These connectors can usually withstand from about 950 to 1,300 pounds of force without allowing the mudsill to slip more than 1/8".

Simpson Strong-Tie "FAP"

Current design of UFP10. Foundation plates should connect directly to the mudsill, as these do.

6.2.1.9.1 Positioning the anchor plates

Foundation plates connect either to the edge or the top of the mudsill. Plywood or mudsill blocks often need to connect to these surfaces; this can cause a conflict between the plywood or block and the anchor plate. So far I have seen about seven different ways to install plywood and anchor plates to the existing construction. Listed from best to worst, these are:

Foundation plate and plywood connected directly to mudsill; plywood cut to fit around plate, and nailed to a single mudsill block along cut edge.

Foundation plate connected directly to mudsill, plywood connected using mudsill block method (shown in middle photo, Figure 6-14)

Installing the anchor plates in the preceding locations would provide the expected loading capacity. Using any of the following locations will degrade the connectors' capacity, some more severely than others.

KC Metals "RFA6." This connector attaches to the top of the mudsill with nails.

Figure 6-14 Several "Foundation plate" connectors.

Foundation plate installed on top of plywood. The left-hand photo in Figure 6-15 shows a UFP10 that was installed after the plywood was nailed onto the mudsill. The UFP10 does not connect directly to the mudsill. This method is barely acceptable. Under the very best circumstances the plywood prevents direct load transfer to the mudsill. Under the worst circumstances the plywood tries to move in different directions from the mudsill and anchor plate as the earthquake forces shift the wall back and forth. In some theoretical cases, loads could travel from the plywood into the anchor plate and then the foundation, and bypass the mudsill altogether—but this would require plywood "edge-nailing" connections to anchor plates. In addition, the plywood moves in different directions from what the anchor plates are tested to resist.

Plywood cut out around foundation plate. This method is not acceptable, as it creates gaps in the plywood nailing and the notches (and possibly over-cuts) weaken the plywood.

Foundation plate connected to mudsill block (either through plywood or directly). The right-hand photo in Figure 6-15 shows a UFP10 connected to a mudsill block behind the plywood bracing panel. This method is not acceptable either; all of the anchor capacity is delivered to a single block with no reliable connections to the adjacent blocks.

Figure 6-15   Foundation plates should be installed directly to the mudsill, as shown in Figure 6-14.   Both of the above installations should be corrected.
Left: Installing UFP10s on top of the plywood is barely acceptable.
Right: This UFP10 connects to a mudsill block behind the plywood. (The arrow marked on the plywood indicates a cripple wall stud visible through the gap at the bottom of the plywood panel)

Foundation plate installed with a wood shim between the plate and a recessed mudsill. The top photo in Figure 6-16 shows a Simpson "FAP" that was installed with a shim. This is not acceptable unless the shim has been securely attached to the mudsill and is good enough quality to not split under loads (such as engineered lumber).

Foundation plate installed with a wood shim between the plate and the foundation. The "FAP" shown in Figure 6-16 is installed over the structural equivalent of a sponge. Sometimes you just need to devise a different connection method; this is one of those times. The "side-bolt" connection is a good alternative to the installation shown—see Figure 7-10

Simpson "FAP" spaced away from mud sill. Shims reduce connection capacity even if they are not split. A "UFP10" is better suited for this connection.

Wood spacers between the foundation and the foundation plate are completely unacceptable. Photo by Bret Butler, GGASHI

Figure 6-16 The above photos show two bad ways to install foundation plates.

6.2.1.9.2 Installing the plates

Installing the foundation plates to the wood framing first allows you to use them as a drilling guide for the holes into the footing. Figure 6-17 shows a UFP10 attached to a mudsill, with holes drilled in the concrete block wall and ready for installation of masonry anchors.

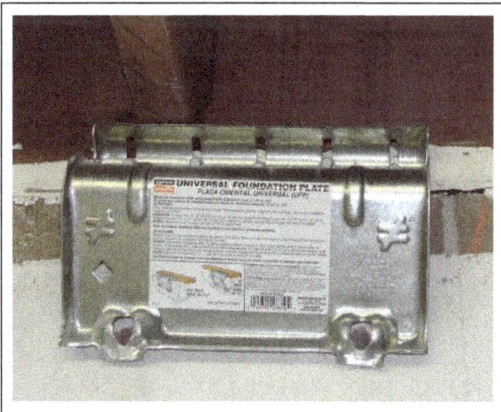

Figure 6-17 UFP10 installed to the mudsill first was used as a template to drill holes in the concrete block wall. Don't forget to install the concrete anchors!

6.2.2 Cripple walls

For houses that have them, cripple walls are by far the most severe weakness in older construction. Figure 6-18 shows a typical cripple wall exposed during a repair project. Usually you will not get to see the outside face of the cripple wall framing. If the siding was painted with lead-based paint, removing the siding requires special lead-containment and disposal measures by trained workers—therefore removing old siding can be more trouble than working inside the crawlspace.

Figure 6-18 Typical cripple wall exposed by removing the lower few siding boards. Photo by Tom Mitchell, GGASHI

Installing plywood panels on cripple walls would be relatively straightforward if they were constructed like a typical wood-framed wall, with consistently sized framing members. At one time houses were built on redwood boards that rested directly on the ground (or mud, hence the term "mudsill"). To keep from sinking into soft ground, wider mudsills were used. Old ways die hard.

Lead-based paints were banned in the US starting in 1978. Many manufacturers stopped using lead in their paints long before that, so houses built in the 1960's may not have lead-based paint. Test kits are inexpensive. EPA fines are not.

Many older buildings in the SF Bay Area have 2x4 cripple wall studs set on a 2x6 mudsill (in multi-story buildings 2x6 studs were often used with 2x8 mudsills). In this case, the interior face of the mudsill is not flush with the face of the studs. Figure 6-19 shows this condition, and how it was incorrectly addressed (in a different house) by toe-nailing the bottom edge of the plywood to the mudsill. We need to create a uniform nailing surface to connect each plywood edge. There are three ways to address this problem, as discussed in the following sections.

Figure 6-19 Mudsills in the Bay Area are commonly wider than the cripple studs. This prevents easy connection of plywood along the bottom edge; toe-nailing (as shown in the right-hand photo) is not an acceptable connection.
Photos: Left—author; Right—Matt Cantor, GGASHI

6.2.2.1 Flush-cut mudsill

Most carpenters in the trades today learned to build walls with material that is all the same width; usually 2x4 stock is used for sole plates, studs, and top plates. By sawing the mudsill to width in place, the flush-cut method gets you to the point where you can proceed as if you were building a wall with uniform lumber size.

Special saws (used primarily in the flooring trades) allow workers to cut the existing mudsill flush with the face of the studs. Figure 6-20 shows a flush-cutting saw and the result of using it to cut the mudsill back so it is even with the face of the cripple studs. [One company that makes kits to convert a regular circular saw to a flush-cutting saw is Flussa; Crain Tools (1155 Wrigley Way, Milpitas, CA 95035, www.craintools.com) manufactures flush-cutting saws. Another manufacturer is fighting an uphill battle to bring a flush-cutting saw to market for general use: www.straightflushsaw.com/]

Figure 6-20 Flush-cut sill method. *Left*: Flush-cutting saw (held backward for the camera). Photo courtesy of Bay Area Retrofit, Inc.
Center: Mudsill cut flush with face of studs. *Right*: Plywood installed

Sometimes mudsills were embedded into the top of the footing while the concrete was still wet. In this case there may be a lip of concrete next to the mudsill, as shown in Figure 6-21. This can be broken off using an air chisel (commonly used for auto-body work).

The wood you will typically find in an old mudsill is high-quality wood, and the lumber is usually a full 2 inches thick. This material resists splitting very well when shear panels are nailed to it.

Figure 6-21 *Left*: Mudsills were often embedded into wet concrete, which obstructs nailing plywood to the face of the sill.
Right: An air-chisel makes short work of removing the narrow fin of concrete

Engineers often express concern about the strength of redwood compared to Douglas fir. This would be a more valid concern when comparing "regular" redwood to Douglas fir—but the density of *old-growth* redwood (listed in the National Design Specification for Wood as "close-grain redwood") gives it almost the same connection capacity as Douglas fir. When anchoring original mudsills to foundations, the full 2-inch thick material gives a higher-capacity connection than 1½" thick Douglas fir.

If you use the flush-cut method you will need to take the narrower mudsill width into account when installing new mudsill anchors.

The flush-cut method allows installing shear walls that most closely match walls that have undergone extensive testing for the last 60 years. This method also exposes some of the inner portion of the mudsill, which offers a chance to evaluate the condition of the wood. In some cases the prolonged contact with damp concrete footings leads to decay of the inner "core" of the mudsill (see Figure 6-22).

I have heard several objections to the flush-cut method; I address these at length in Appendix D. In my opinion there is rarely a valid engineering concern about using the flush-cut method.

Figure 6-22 The underside of a 100-year old mudsill, exposed during a foundation replacement project. Even foundation grade redwood begins to decay when exposed to dampness for decades.

6.2.2.2 Reverse block

The reverse block method is illustrated in Figure 6-23. In this method you attach a length of lumber (the "reverse block," which may be several feet long) to the shear panel before installing the shear panels to the studs. The panels are nailed to the "back" of the reverse block. After installing the panels with the pre-attached block, the block is nailed to the existing mudsill.

In some cases existing mudsill anchors will obstruct installation of the reverse block. While it is best to remove the obstructions, it is not always easy to do and you may have to notch the reverse block or shear panels to fit around them. Notching the reverse block is rarely a cause for concern, because very little force builds up in the reverse block (loads from the plywood panels transfer through the nearby nails from the reverse block into the mudsill).

Figure 6-23 Reverse Block Method viewed from front of wall (shown at left) and back.. Three plywood nails visible in the left-hand photo along the bottom edge go into the edge of the flat 2x4 reverse-block in the right-hand photo. The 2x4 reverse-block is nailed to the original mudsill. New expansion anchor securing mudsill to footing shown in right-hand photo.

The reverse block method has a couple of drawbacks. First, in most cases there is no way to verify the plywood attachment to the reverse block once it is installed (the wall shown in the example is between a garage and a crawlspace, where no exterior sheathing obstructed viewing the back of the plywood). Second, care is needed to assure that the reverse block will bear solidly on the mudsill (sometimes the original mudsill is not flat). The method also requires prefabricating the reverse block and plywood assembly, which presents an opportunity to assemble pieces backwards or upside down, possibly to discover your mistake only after you have dragged them into the crawlspace.

6.2.2.3 Mudsill blocking

In the mudsill blocking method, new blocks are installed on top of the mudsill between the cripple studs, with their edges flush with the cripple studs as shown in the left photo in Figure 6-24. This provides a nailing surface along the bottom edge of the plywood. A major concern with mudsill blocks is that the blocks tend to split, as short pieces of lumber split much more easily than longer pieces (see Figure 6-25). One way to decrease the chance that mudsill blocks will split is to cut the bottom off one or two studs just enough to slip the block under them, in which case you can increase the block length to match the width of two or even three stud bays, as shown in the drawing in Figure 6-24.

Figure 6-24 Mudsill blocking; *Left:* Block shown at left is ready to fasten to original mudsill (the right end should be nudged forward about ¼" so it is flush with the face of stud at right). *Right:* A mudsill block spanning two stud bays to help keep it from splitting when it is connected to the mudsill and receives plywood edge nailing. Photos: Left—author; Right—Kevin Farrell

In my opinion there are too many opportunities for installation errors in the mudsill block method. I much prefer the flush-cut method; currently there are very few contractors who have the hard-to-find tools required.

If you use mudsill blocks you need to avoid obstructions such as pipes or wires that pass through the mudsill. Blocks must be notched to fit over existing mudsill anchors; new anchors should be installed through the blocks for the simplest connection.

Connections are crucial (as usual). It is important to remember that the block, mudsill, and foundation must each be connected to the adjacent member. That is to say, the blocks must connect to the mudsill, and the mudsill to the foundation. Hoping that a foundation anchor will connect a mudsill block to the foundation without making the crucial block-to-mudsill connection creates an unsecured shear plane that ruins the overall load transfer system.

There are three sub-categories of the mudsill block method based on fasteners used to connect the block to the mudsill: Nails, screws, or structural staples.

Nailing blocks to the mudsill between the cripple studs so the blocks are flush with the studs gives you an even surface to attach the shear panels. This method was developed to allow homeowners to retrofit their own houses, as it is relatively simple and does not require special tools like the flush-cut method does. But this method has some problems, as follows.

First, testing has shown that the blocks tend to lift up off of the mudsill during side-to-side loading. This would be expected unless the blocks were installed underneath the nut and washer for new mudsill anchors.

Second, nailing the blocks to the mudsill often results in the blocks splitting. Split blocks have almost no connection capacity. Even if you check the blocks and see no splitting before you install the plywood panels, you are not finished driving nails into the blocks. All the panel edge nailing into the blocks could cause splitting—and in most cases you will have no way to inspect the blocks once the panels are in place (usually you are installing plywood on the exterior cripple walls).

Homeowners might take the time to pre-drill the blocks before nailing them in place. Contractors need faster installation methods.

Using 3x4 material for the blocks will greatly reduce splitting problems. However, now you will need extra-long nails to connect the 2½" thick blocks, and longer foundation anchors as shown in Figure 6-26.

Of several reviewers and contributors asked specifically about the nailed-block method, none of them recommend it over the flush-cut method. I suggest that only homeowners use the nailed-block method, and that they keep in mind that their home depends on careful installation.

Figure 6-25 Mudsill blocks that split when nailed to the mudsill. These blocks were installed on interior cripple walls; usually you are bracing perimeter walls and the blocks are completely concealed after installation. In that case you cannot tell if they split when you nail the plywood to them.
Photos: Paul Rude, GGASHI

Attaching blocks using structural screws reduces splitting the wood. When I need to use mudsill blocks I specify attaching them with self-drilling structural screws, as shown in Figure 6-26 and Figure 6-27. Various screws with high connection capacities include Simpson's "SDS" and "SDWS" series, USP's "WS" series, or Fasten Master's "Trusslok" series. Attachment using screws also reduces the chance of the blocks lifting off the mudsill.

One of the few conditions where the mudsill-block method is essentially the only choice is when the existing mudsill was embedded into a concrete floor slab, making it impossible to chip away the obstructing concrete as shown in Figure 6-27 for the flush-cut method. If there are other obstructions that prevent using the reverse-block method (say the mudsill is set slightly below the surface of the concrete) then mudsill blocks are warranted. Use sill anchors that secure the blocks to the footing, or else install tie-downs when feasible, and use nails only as a last resort to attach the blocks.

A final attachment method is air-driven structural staples. Staples are small enough that they do not split the blocks very easily. Even though staples have legitimate structural ratings, they are often looked at as inferior. Adding a few extra staples is easy; this is good assurance that the staples will not fatigue and snap off at the wood joint, as has been shown to happen when they are used to attach plywood bracing panels.

Figure 6-26 Mudsill blocks made from 4x4 stock. The foundation anchors installed through the blocks will help keep them from lifting off the mudsill during an earthquake.

Figure 6-27 Mudsill block attached to the mudsill with structural screws. In this situation mudsill blocks are needed even if the mudsill and studs are the same width, because a garage floor slab was poured up to the top of the mudsill. This installation will be completed by adding foundation anchors with nuts and square washers.

6.2.2.4 For short walls: Match the mudsill by building out the top plate and end studs

When building shear walls, the wall studs can be spaced up to 24 inches apart. Plywood does not know which way is up—all it knows is that it has to be attached to framing members along each edge. If you take a 2-foot wide, 8-foot tall shear wall and roll it a quarter turn you end up with an 8-foot wide, 2-foot tall shear wall. The important distinction is that this wall has no intermediate studs; to resist earthquake forces only, it does not need studs.

In cases where the cripple studs are less than 22 inches tall, you can simply add members to build the framing out to the width of the mudsill. You will need new backing behind all panel edges; for 8-foot plywood panels installed with the long dimension horizontal, you need to build out a stud (or install a new post) only every 8 feet. Sometimes the thickness you need to make up will be 1½" and you can use a 2x4 flat against the original framing. Often you need to build up a full 2"—in this case I recommend installing a flat 2x4 topped with a layer of ½" plywood that is glued to the flat 2x4 (see notes about adhesives in Section 4.3).

6.2.2.5 Venting for stud cavities

Most jurisdictions require vent holes in shear panels that would otherwise enclose stud bays in a crawlspace. The minimum vent hole size in the Bay Area is 2½" diameter. A single hole is allowed at mid-height of stud bays up to 18" tall (see the middle photo in Figure 6-14). For taller cripple walls, bore one hole between 3" and 5" from the bottom of the panel, and another at the same spacing from the top, as shown in Figure 6-28.

When aligned with mudsill anchor bolts, vent holes allow for inspection of the bolts.

Some jurisdictions (all of the City of LA) require installing ¼" mesh over the vent holes to keep rodents out (see Figure 6-26; note that this contractor installs screens on all holes before hauling the plywood under the house, even when a rodent—or an engineer—can go around to the back of the panel).

In flood-prone areas the code prohibits any construction that would cause water to build up on one side of a wall. Vent holes allow pressure from floodwaters to equalize.

Vent holes should be cut with hole-saws or appropriate auger bits—do not plunge-cut holes with a circular saw. The triangular vent openings shown in Figure 6-29 weaken

Figure 6-28 Vent holes bored at top and bottom of plywood panels.

Figure 6-29 BAD—Was this wall a practice-run for carving pumpkins? Cutouts with sharp corners encourage the plywood to tear.
Photo: Paul Barraza, GGASHI

the shear panels and scream "amateur installation."

6.2.3 Use borate-treated lumber for any new pressure-treated lumber

In some cases you will need to install new foundations or mudsills, or both. Sometimes studs or posts are placed against stepped foundations; wood in contact with concrete must be treated to resist decay (or be naturally decay-resistant) whether the member is oriented horizontally or vertically. Figure 6-30 shows a severely rusted anchor bolt used in an earthquake retrofit. Rust is the simple term for "corrosion," which is the chemical process in which different metals react in the presence of water. Moisture found in crawlspaces—even ones that do not seem "damp"—is enough water to cause this reaction. See Section 8.5 for information on corrosion problems with PT lumber. Home inspectors in the SF Bay Area are growing increasingly concerned about new lumber treatments other than borates, and will often cite ACQ or other copper-based chemicals as a deficiency even if it meets the building code. You do not want to be the person who installed questionable materials because that's all they had at the big box store.

I specify only borate-treated lumber on my projects. If I could use flashing text to note this on plans, I would do so. Borate-treated lumber is usually a special-order item; **borate-treated wood is also usually less expensive** than the copper-containing treatments. In some cases homeowners or contractors unfamiliar with my level of concern over this will install "regular" pressure-treated lumber. If I see copper-containing treated wood on projects I have designed, I require that it be removed and replaced with borate-treated lumber, or that stainless steel connectors and fasteners be used in contact with the treated lumber. If I am evaluating existing construction that has treated lumber I also specify stainless steel connectors. This is something that I check for on **every** project, and I will **always** specify stainless steel if I see copper-containing wood.

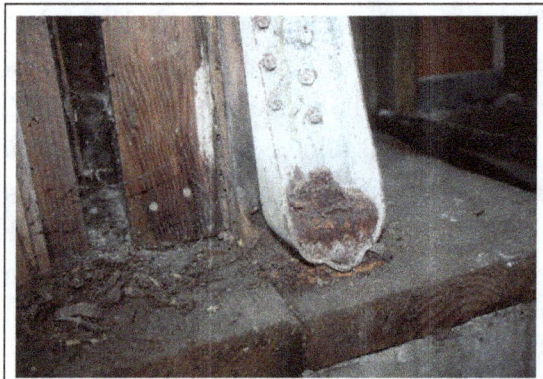

Figure 6-30 The chemicals used since 2004 (and some extent before that) to treat wood for decay-resistance also lead to aggressive rusting of steel fasteners. Severe rusting like that shown above can occur in just a few years. THIS IS A VERY SERIOUS PROBLEM. See Section 8.5 for more information.
Photo by Paul Rude, GGASHI

6.2.4 Connecting the Floor Diaphragm to the Structure Below

Earthquake forces generated in the structure above the crawlspace or basement/garage areas are generally spread out through the floor diaphragm. In most cases the diaphragm sheathing is connected to the floor framing members adequately to transfer forces into the floor framing, and we need only to supplement the floor framing connection to our braced walls. In some cases the floor framing probably has adequate connections to the braced walls as well. This section discusses a variety of situations and several ways to supplement the connections.

6.2.4.1 Framing connectors, aka Shear Transfer Ties, aka "STTs"

Various sheet-metal connectors work well to attach joists or blocking to the mudsill or cripple wall top plate. The following connectors are rated for significant load resistance and install relatively easily with a palm-nailer.

LTP4 & LTP5 for flush surfaces. When the block or joist surface is flush with the framing below, flat connectors like the LTP4 work well. Figure 6-31 shows a row of Simpson "LTP4" connectors attached along the split-level floor at the rear of a garage. Plywood will be installed on the cripple wall below the split-level.

Both the LTP4 and LTP5 use 8-penny nails, or preferably Simpson's 1½" long "N8" nails made specifically for use with sheet metal connectors.

Figure 6-31   Shear Transfer Ties connect from joist blocks to the cripple wall top plate (studs were temporarily removed for access to the crawlspace beyond).

Angle-type connectors work well when surfaces are not flush. KC Metals' "CA" series and Simpson's "L" series of connectors are intended to join members that meet at 90 degrees, such as the case in Figure 6-32 where a joist is inset from the face of a shear wall.

These connectors accept 10-penny nails. Some engineers do not like the "CA" or "L" series of connectors because the 10-penny nails are more likely to split the wood than 8-penny nails used in other connectors. In my opinion the advantage of nail placement outweighs this—and you can always use 8-penny nails if you want (with a reduction in the connection capacity, which is usually in fine print in the manufacturer's catalog). Simpson recently

Figure 6-32   A "CA90" connector from KC Metals connects the top of a braced wall to the joist above it.

introduced self-drilling structural screws that they have tested with these connectors; using these special screws would likely reduce splitting as well.

Driving more than three nails into a short block risks splitting the wood. I specify short angle connectors like Simpsons "L50" or KC Metals' "CA50" for connecting blocks. For continuous joists the connectors that use more nails generally work fine. In any case, using a full-size 10-penny nail (0.148" diameter, 3" long) is impractical; Simpson's 1½"-long "N10" or similar nails are a better choice. In some cases shorter nails are essential because a full 3-inch nail is impossible to install, or, worse yet, would poke out through the siding.

Installing shear transfer ties is almost impossible without a palm-nailer. The left-hand photo in Figure 6-33 shows an L90 peeking out from a joist space no more than 6 inches wide. Using a hammer to drive nails in such spaces is insanity. Palm-nailers fit into such spaces fairly easily—the model pictured is a low-end tool, but also extra small.

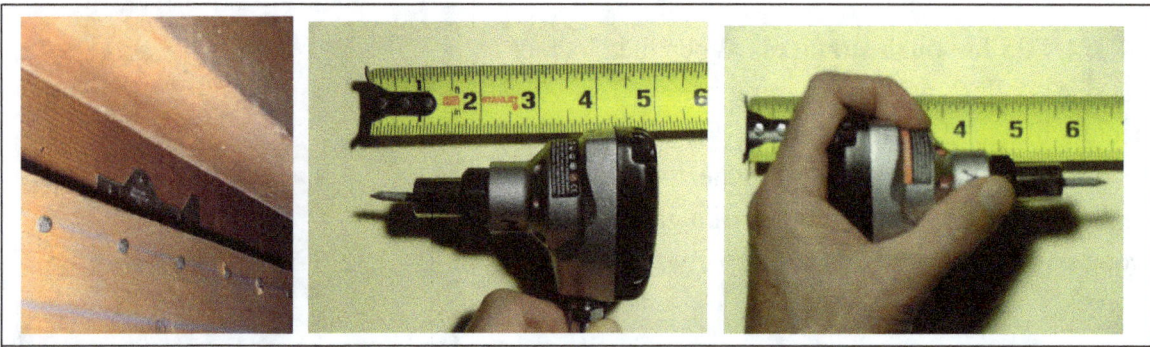

Figure 6-33   Left: When it's hard even to photograph the hardware, installing it must have been tough. The other photos show a palm-nailer that fits in tight spaces.

6.2.4.2 "H10AR" and similar connectors

Figure 6-34 shows a Simpson H10AR used to transfer shear forces from the joist to the top of a shear wall. Simpson lists this as a "hurricane tie," but unlike most hurricane ties—which are essentially useless for resisting horizontal earthquake forces—this connector secures both sides of the joist. It is rated for 590 pounds acting along the top of the wall (compared to 100 to 135 pounds for a typical hurricane tie).

Joist blocking is essential when using H10ARs.
The H10AR and similar connectors rely on lateral forces transferring from the floor diaphragm nailing at the top of the joist to the bottom portion that is gripped between the connector flanges. Blocking is required on both sides of the joist to provide a strong enough load path from the top of the joist to the H10AR. If no blocking is present (or is present on only one side of the joist has a block) then the opposing forces on the joist cause what engineers call "cross-grain tension" or "cross-grain bending." Wood is extremely weak in resisting cross-grain bending—so weak that the building code tells designers to avoid connections that rely on cross-grain bending. **Connectors such as the H10AR should be used only on joists that are stabilized with blocking on both sides.** Joists that are end-nailed from a rim joist will not have sufficient resistance to cross-grain bending. It is not always easy to tell whether you are looking at blocking between joists or a rim-joist that they butt into; if there is any doubt, use an alternative shear transfer tie, such as the LTP4 or L50.

Figure 6-34 H10A or H10AR are about the only "hurricane ties" suitable for earthquake retrofit work. (Per the manufacturer, the gap between the top tabs of the tie and the joist blocks is acceptable.) Plywood will be installed over the tie where it attaches to the cripple wall top plate.

The current installation instructions for the H10A do not require nailing to the blocking on either side of the joist. I note this because inspectors sometimes want to see nails installed in every hole in a connector. The H10AR in Figure 6-34 is acceptable as installed, with nails to only the joist and wall plates. (This could change; you should check the instructions in your hardware manufacturer's catalog).

6.2.4.3 Structural screws up through the cripple wall plate

Self-drilling structural screws can be installed up through a cripple wall top plate into joist blocking or a rim joist, as shown in Figure 6-35. In finished or semi-finished rooms this can eliminate a great deal of work required for access to install a sheet metal connector, plus patching afterward. Take appropriate precautions to assure you do not drill into existing utilities.

Figure 6-35 Structural screws (circled in yellow) driven up through the cripple wall top plate into joist blocks above will serve to transfer forces from the floor framing to the braced wall.

6.2.4.4 New joist blocking using sawn or engineered lumber

If the original construction did not include joist blocking you may need to install new blocks. Blocks must serve as "shear elements"; that is, they must transfer forces from top to bottom and side to side (for more information refer to the *Shear Wall Guide*). This means all four sides should connect to framing—and if the block splits (see Figure 4-7) it no longer functions as needed.

Only kiln-dried lumber or engineered wood (including plywood) should be used for blocking in existing construction. The edges of the blocks are necessarily fastened to the existing structure, essentially forcing the blocks to split as they dry out.

Blocks should fit snugly between the joists. Joist bays are seldom the same width, and are often not rectangular; this requires careful measuring and cutting. This typically means toe-nails or end-nails at both ends and attaching the top with construction adhesive (typically used for attaching sub-flooring in new construction) to the underside of the floor sheathing. Once installed, the blocks can be connected to the framing below with any of the methods listed previously.

One contractor I work with estimates block installation at $50 apiece. This adds up quickly, and engineers should consider just how many blocks are really needed to affordably transfer earthquake forces.

6.2.4.5 New blocking using plywood "blocks"

Some retrofit guidelines recommend using ¾" plywood cut to fit snugly between the joists if blocks were not installed originally. Plywood is cut in pieces taller than the joists so it can lap onto the mudsill or cripple-wall top plate; the plywood blocks are face-nailed to the mudsill or plate. This method does not provide for positive connection between the top or sides of the block. Although a snugly-installed block will transfer some force just by virtue of bearing against the joists, it is best to connect it along all edges. Attaching plywood using toe-nails is a very bad practice, even for plywood as thick as ¾". I recommend connecting with a generous bead of construction adhesive.

6.2.4.6 "Lock-blocks"

This method was developed by a veteran retrofit contractor to address the time-consuming measuring and cutting needed to fit a single block between joists. Instead of a single piece of plywood, two pieces that are each slightly narrower than the joist bay are used. One piece is installed against the joist on the left, and another is installed on top of the first piece and against the joist on the right, as shown in Figure 6-36. Both pieces are face-nailed to the mudsill or top plate. Again, I recommend attaching the edges of the blocks with construction adhesive—and also connecting the two plywood layers together.

Figure 6-36 "Lock-blocks" installed in segments glued in place to fit tightly between joists. This method eliminates a lot of fussy measuring and cutting. In this project the lock-blocks are attached to a collector that will transfer forces to a braced wall at right.

6.2.4.7 In some cases, shear transfer ties may be redundant

Section 4.1.2.3 discusses the original connection between the joists and the framing supporting them. For single-story houses the original connections are likely adequate, provided that blocking was installed to stabilize the joists (see Figure 4-5). For two-story houses the original joist connection is probably adequate where the walls perpendicular to the joists are significantly longer (at least 50% longer) than the parallel walls—again assuming that blocking is present. In either case, if blocking was not installed originally you must provide it as part of the retrofit.

6.2.5 Collectors & ties

Collectors and drag-struts (or drag-ties) transfer lateral forces from one area of a building to another. Forces need to transfer from one part of a braced wall to another unless you cover the *entire* length of the wall with plywood. Drag ties are also often needed at inside corners of the building floor plan. Collectors and ties can get complicated pretty quickly. The *Shear Wall Guide* devotes over 12 pages to collectors. Proper collector and tie design is best left to an experienced engineer.

All I will address here is the most basic collector: wall top plates. Since wall top plates also perform another purpose as diaphragm "chords" they need to be strengthened for two reasons.

Figure 6-37 GOOD: The joint in this cripple wall double plate is spliced with dozens of nails connecting the double members.

Installing shear panels on the entire length of existing cripple walls is usually impossible. The cripple wall top plate connects a long length of floor diaphragm to the shear panels installed where they will fit. In such cases the top plates "collect" the earthquake force in the floor and "drag" it to the shear wall segments. If there are no breaks in the top plates we usually have nothing to worry about. Typically, though, there are some joints in the top plates. The joints must be supplemented with additional nails on both sides, as shown in Figure 6-37.

Determining just how many nails to install can become a complex engineering exercise, and this book is not meant to teach engineering. You could hire an engineer to determine whether you need 27 nails or 37 nails on each side of a particular joint, but it's probably more economical just to install the maximum number that makes sense. At some point the strength of the top plate becomes the over-riding factor. A full-dimension 2x4 has a cross-sectional area of eight square inches and (assuming old-growth fir) has a tensile strength of about 1,200 pounds per square inch. If our 2x4 has not decayed or been munched by termites, notched to accommodate pipes, drilled for electrical wires, or otherwise insulted, it has a capacity of almost 10,000 pounds in tension. (Note: For most houses, this is *much* more force than we would calculate.) To develop 10,000 pounds of force requires lots of nails or other fasteners. Assuming a full 2-inch thick member we need to use 16-penny common nails—but how many?

Warning: Math.

> 16-penny nails resist about 220 pounds of earthquake force each.
>
> 10,000lbs/220lbs/nail=45 nails

If you know the strength of other fasteners such as GRK or Fasten-Master screws, lag screws, etc. you can calculate the "maximum reasonable number" of them to use.

If we have a 1¾" x 3¾" net member (common in the SF Bay Area in the early 1900s) we can use 16-penny sinkers instead of commons. By coincidence we still need 45 nails; the lower member strength of the smaller lumber size corresponds to lower capacity of the thinner nails.

The nails must be well spaced and staggered to prevent splitting. Pre-drilling for the nails with a drill that is about three-fourths of the nail diameter would also help prevent splitting.

An alternative to nailing or screwing the double plates together is installing a steel strap across the splice in the plate, as shown in Figure 6-38. Again, finding the "correct" strap capacity involves engineering. The most important thing is to install the strap properly.

Since collectors and ties almost always have to act in tension or compression you will have to fill any gaps in members joined with straps. A strap installed across the 1-inch gap shown in Figure 6-37 could withstand tension, but the strap would buckle if it was subjected to compression forces.

Figure 6-38 Joint in top plate spliced with a steel strap to tie the wall at left to the newly-braced wall at right.

The photos in Figure 6-39 and Figure 6-40 show some other instances of collectors and ties. If you are evaluating an existing retrofit installation and come across this sort of thing, it probably has a purpose. If you are a plumber, don't cut it out of the way. Please.

The fireplace shown at left above interrupts the cripple wall. The 4x4 strut shown in the right-hand photo ties the portion of the floor framing in the far background to a new section of braced wall to the right (outside of the photo). The 2x6 was installed to hold the 4x4 away from the masonry. This is a "tie" since it is not collecting forces along its length.

Left: The plywood lock-blocks between the joists transfer force from the floor diaphragm to the 4x4 strut that butts into the mudsill in the background. *Right*: close-up of connection from collector to mudsill. Forces transfer from the mudsill to the foundation through the UFP10s.

Figure 6-39   Other examples of drag-ties and collectors.

Left: The beam at left collects earthquake forces; the pair of brackets (usually used as "hold-downs," but we love them for other things, too) and all-thread rod tie the beam to another beam behind the plywood panels. Note: Whenever possible, mount hardware directly to the framing—not through the plywood as shown. Also keep the brackets and all-thread as close to horizontal orientation as possible.

Right: Another pair of brackets used with all-thread rod between them to splice the top plate (actually a 4x6 beam) together where a floor girder runs through.

Left: The steel strap ties a porch beam to the new 4x member. *Right*: Existing joists joined together where they lap over a floor girder; use care to space fasteners far enough apart so that they do not split the wood. Some manufacturers list spacing requirements in their instructions.

Figure 6-40   Still more examples of drag-ties and collectors.

6.3 Nonstandard Construction Conditions

Many framing conditions are outside the conditions shown in the IEBC or most other standard retrofit plans provided to date. Some conditions are simply due to the original building dimensions or building customs; others may have been constructed by early tribes of surfers.

6.3.1 Joist spacing too close for access to end joist

In order to connect the end joist to the top of a cripple wall or mudsill, you either need enough room to reach up above the cripple wall top plate or enough clearance to install screws up through the plate(s) into the end joist. In some cases the framing layout is such that the last joists are only inches apart, and you cannot install an L90 or the like; Figure 6-41 shows such a condition. In this case you need to connect to the next joist in from the end and connect that joist to the shear wall or footing with a "shear transfer diaphragm" (see Section 6.4.7). If the second joist is close enough, a "side-bolt" connection from the footing with framing connectors to the second joist can work, similar to that shown in Figure 7-10.

Figure 6-41 The joist spacing shown does not allow working room to connect from the end joist to the mudsill.

6.3.2 Balloon framing

Balloon framing is the term used for walls constructed with studs that run continuously from the mudsill up past one or more floor levels (see Figure 6-42). Floor joists are nailed to the faces of the studs, and may also be supported on a ledger or let-in ribbon. There is no top plate in the crawlspace for attachment of shear panels. There are also combinations of standard "platform" framing and balloon framing.

In any framing scheme, earthquake forces must transfer from the floor diaphragm into the shear walls. Balloon framing almost always requires installing blocks in at least one location, as shown in Figure 7-18 and Figure 7-19

Figure 6-42 Balloon framing; studs run past the floor joists. Where joists run into the wall, they are supported on the 2x6 ledger and face-nailed to the studs, as shown on the left side of the photo. The end-joist running the same direction as the wall is nailed to the edges of the studs. Gaps between the studs should have "fire-blocking" that will prevent flames from spreading through the wall cavities.

> **Balloon framing** minimizes the change in building height due to wood shrinkage. Platform framing can shrink as much as an inch per floor level. Balloon framing is still used to frame houses that will have masonry veneer attached (a bad idea in seismic areas). The building code still references balloon framing connections.

6.3.3 Floor sheathing connected directly to mudsills

On sloping lots, the subfloor sheathing may attach directly to the mudsill along the uphill edge of the floor platform (see Figure 6-43). In this case it is often impossible to install additional mudsill anchors. Connecting the first joist to the footing with a shear transfer diaphragm, as shown in Figure 7-24, usually works in such cases.

Figure 6-43   The subfloor connects directly to a mudsill on top of the foundation wall at right. Supplementing the mudsill connection to the foundation is sometimes impractical in this case.

6.3.4 "Hip and shoulder" footings

Another construction method used on sloping lots is a two-level footing, where one mudsill supports the floor framing members; a "curb" extends higher and supports another mudsill to which the subfloor attaches. Such a connection and one retrofit method is shown in Figure 6-44. Hip and shoulder footings use several possible positions of joist blocking, which affects what connection is best; see Figure 7-12 for another common configuration.

Figure 6-44   *Left*: This "hip and shoulder" footing supports the floor joists on the lower mudsill and the sheathing on the upper mudsill. The upper mudsill is hidden behind tar-paper, just below the scorch marks and holes in the plywood.)
Right: One way to connect from joists to mudsill uses "H10AR". A "UFP10" completes the load path from mudsill to foundation.

6.3.5 Cripple walls with no top plates

One early "budget" construction method was to support the end joist directly on top of the cripple studs without using a top plate, as shown in Figure 6-45. In this case there is no nailing surface for the top edge of shear plywood. Figure 6-46 shows the gap in the load path that results if you ignore this condition.

Figure 6-45 Cripple wall with no top plate offers no place to nail the top edge of plywood bracing panels.	Figure 6-46 **WRONG way to in-stall plywood if there is no top plate.** (Photo taken from inside the joist bay looking down—there is an original stud just to the right of the bunch of twine, with shadow on it)	Figure 6-47 Sometimes carpenters tried to balance framing members on a single cripple stud. At least this joist does not have far to fall.

Installing an additional layer of framing lumber to the inside face of the end joist can usually provide a flush surface for plywood connection, as shown in Figure 7-16. Often a new 2x member is just the right thickness; you can add plywood if you need to (with plenty of construction adhesive). Nail the new member to the existing end joist with 16-penny nails at the same spacing as your plywood edge nail spacing—slightly more if you are using 10-penny plywood nails.

The framing condition shown in Figure 6-47 is extremely rare. In this case, joists perpendicular to the wall are each supported atop a single cripple stud, sharing the bearing area with the rim joist. This could be solved with a combination of methods, none of them easy.

6.3.6 Studs overhang the mudsill

Figure 6-48 shows an interior cripple-wall bearing to one side of the footing. Where the studs are offset from the mudsill you have very little room to install mudsill blocks, and if you cut the mudsill flush there will not be enough of it left to provide for plywood nailing nor installation of new sill anchors. In this case you can often use the reverse-block method to connect the bottom edge of the shear panels. Another option is to install new studs that land fully on the mudsill and make appropriate connections top and bottom.

Note that this condition is an original defect in the construction because it does not provide for the intended bearing area of the studs on the mudsill, and it loads the foundation off center. Failure in an earthquake is extremely unlikely. This book is intended to instruct on how to retrofit your house affordably, not on how to fix everything wrong with it. Contractors bid replacement footings at about $600 per running foot—money probably best spent elsewhere.

Figure 6-48 Original construction with studs overhanging the mudsill. There are several ways to deal with this; the most expensive is to replace the foundation.

6.3.7 Mudsill overhangs the footing

The left-hand photo in Figure 6-49 shows a mudsill that does not bear on the footing at all (it is supported on piers about 8 feet apart). To provide a shear connection to the foundation, a new treated member was installed to the side of the stemwall.

Figure 6-49 *Left*: Mudsill viewed from below. Red arrow shows overhang from foundation; black arrow shows light shining through gap at footing. *Right*: Viewed from above, with a new PT 4x4 bolted to the side of the foundation, and the original mudsill attached to the new one with galvanized structural screws. Plywood is not installed yet.

6.3.8 Stacked mudsills

Sometimes more than one layer of lumber was used as a mudsill, as shown in Figure 6-50. The installation shown has four layers of mudsill; together with the end joist and foundation, these create five separate "slip-planes" that must all be secured to prevent sliding.

If the faces of the members line up, multiple sills can be treated as a very short cripple wall and plywood installed with nailing to the top and bottom members. Another option is to install LTP4s or similar connectors across each joint between members. The installation shown in Figure 6-51 is incomplete—the LTP4 does not connect all the members together (in this instance four separate rows of LTP4s would be needed to connect across each slip-plane; a plywood strip across all five members, and nailed to the top and bottom ones, would be much easier to install).

You may find an offset from one layer to the next, as shown in Figure 6-52 where a 2x6 "starter sill" was placed on top of the concrete, followed by a 2x4 sleeper, followed by the floor framing. The two rows of STTs address this: one row connects the rim joist to the sleeper, and a lower row connects the sleeper to the mudsill. UFP10s complete the load path to the foundation.

Figure 6-50 Layers of mudsills create more places where the members can slide on top of each other.	Figure 6-51 BAD: The installation shown above is INCOMPLETE. Each board must attach to each adjacent member; the upper four pieces are poorly joined, and do not connect to the mudsill.	Figure 6-52 Mudsill layers of different widths joined to each other and to the end joist with KC Metals "CA90" connectors. UFP10s complete the load path to the foundation.

6.3.9 Perimeter foundation that wraps around a porch rather than supporting a building wall

My grandmother's 1906 house had a front porch across most of its width. The porch had a full foundation under the sides and front edge—but the front wall of the house was supported by a row of posts and piers. A new shear wall aligned under the front wall of the house would have no foundation to connect to.

The best solution is to install a new footing in line with the front of the house. This is also quite expensive. Usually there is a double joist supporting the original wall, and your new shear panels can connect to that member. Sometimes this is a good opportunity to re-level a building that has settled.

Figure 6-53   Front wall without foundation. The photo at right was taken in the crawlspace looking toward the front of the house. The foundation follows the porch outline; there is no foundation below the front wall of the house.

Instead of installing a new footing and braced cripple wall under the porch, some contractors install a diagonal brace that connects to the existing footing, as shown in Figure 6-54. I don't recommend this for a few reasons: It places a large load on the original footing, which is probably not in good condition; retrofit anchors securing the end of the brace are often loaded in withdrawal from the old concrete, which I do not view as reliable; and depending on the angle of the brace from horizontal, it could exert overpowering up or down forces on the joist to which it connects. However, this alternative is better than doing nothing at all. I would hope it is viewed as a step toward a better solution, though, and a new footing and shear wall will be installed in the future.

Figure 6-54   This diagonal strut in line with front porch wall and connected to foundation is less effective (but much less expensive) than a new footing and cripple wall braced with plywood.
Photo: Paul Rude, GGASHI

6.4 Existing obstructions

Retrofit guidelines don't usually show the maze of wiring, pipes, sewer lines, plasma conduits, Jeffrey tubes, ducts and so forth that most houses feature. Figure 6-55 shows a sample of what can impede shear panel installation.

Preferably you can avoid installing retrofit work in congested areas. Relocating an electrical panel could easily cost a thousand dollars or more, given that you may need to upgrade to contemporary wiring methods. Likewise, moving an old cast-iron sewer line may develop into an enormous project as more and more of it keeps crumbling away as you try to reconnect it. Designers should do their best to keep new work away from existing utilities, both for installation ease and also the danger of future Sawzall-wielding tradesworkers who only view your retrofit as an obstacle.

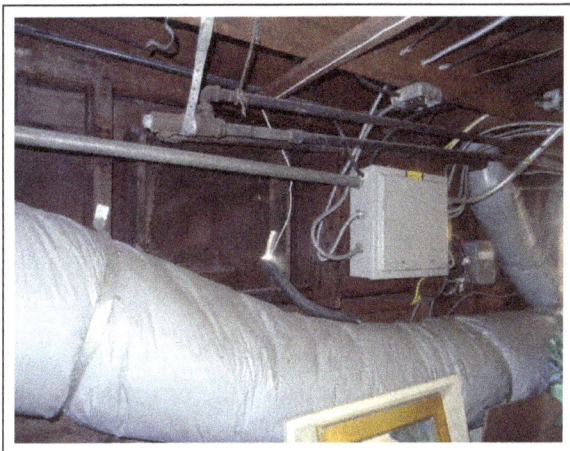

Figure 6-55 "This wasn't in the tour guide!" Eliminating work in congested areas as shown above is an important part of a good retrofit design. Prescriptive designs often do not allow much variance from required plywood panel placement.

Despite your best efforts to avoid utilities, there are plenty of instances where you simply have to work around an obstruction. The following methods address the most common obstructions.

6.4.1 Hazardous wall coverings

Whether you view them as hazardous or not, some wall coverings may be viewed as hazardous by the building department, OSHA, the owner, or the "expert" the owner will meet at a dinner party the day after you tear coverings off their walls. Lead paint and asbestos are the most commonly acknowledged hazards. Removing old painted paneling or asbestos-containing drywall could easily double the cost of a retrofit.

In some cases it would be possible to install plywood on top of the existing wall coverings. The building codes allow installing plywood on top of gypsum wall-board if you increase the size of the nails to the next larger size. This would allow more movement of the braced wall, but may be the most economical choice.

Installing plywood over existing wall coverings requires care to accurately locate the studs beyond, so that your plywood nailing connects to the framing. Consider pre-drilling holes in framing if using 10-penny nails (check with the building jurisdiction first to learn how much hazardous material you can disturb without needing special containment measures).

If you cover gypsum board with plywood, you have changed the fire-resistive condition of the walls; you may need to install another layer of gypsum board over the plywood.

The biggest problem in applying plywood over existing walls is that you cannot access the mudsill to install supplemental anchors. You will need to use a connection method such as the "Side-bolt" or "Reverse block" which will protrude into the finished space somewhat.

6.4.2 Pipes & conduit passing through walls

The very best way to run a pipe or similar through a shear wall is to install the shear panels first, then drill an appropriately sized hole through the panel (at least two inches from any panel edge) and install the pipe. Since chasing down a plumber or electrician could add several hundred dollars to the project cost, not to mention likely delays, this option is seldom chosen.

The next best way to work around a pipe is to arrange your shear panel layout so a panel joint occurs at the pipe or other penetration, and make the smallest possible cut-out for the pipe. Such an installation is shown in Figure 6-56.

Sometimes there are several pipes close by, and you have to slot the plywood to allow installation. Or sometimes you are repairing an installation from an earlier time, as shown in Figure 6-57. Remember: every edge of plywood must connect to some other structural element. In this case, attach a plywood patch over the slot with construction adhesive. The patch should cover the entire length of the slot, and lap 6 inches onto the un-cut shear panel on both sides of the slot. Use enough adhesive to completely bond the two surfaces, and secure the patch in place with small self-drilling screws (drywall screws would be acceptable, as their function is only temporary).

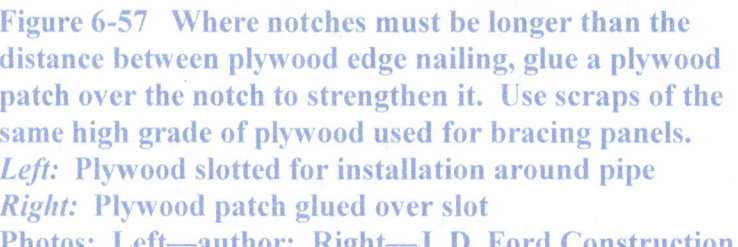

Figure 6-56 Notches in plywood should be as small as possible, and have rounded edges.
Photo: Paul Barraza, GGASHI

Figure 6-57 Where notches must be longer than the distance between plywood edge nailing, glue a plywood patch over the notch to strengthen it. Use scraps of the same high grade of plywood used for bracing panels.
Left: Plywood slotted for installation around pipe
Right: Plywood patch glued over slot
Photos: Left—author; Right—J. D. Ford Construction

> Skeptics of gluing plywood patches in place should note that plywood is manufactured with glue between the layers in the first place. I have had to demolish construction assembled with contemporary adhesives, and I have no doubt that the attachment suggested here is adequate.

6.4.3 Pipes & conduit running along walls

Non-metallic sheathed electrical cables ("Romex") and flexible armored cable running along walls are reasonably easy to gently remove before you install plywood and then reattach afterward. Despite the relative ease, a surprising number of retrofit installers do not want to move such obstructions, and install plywood stopping just short of where it needs to reach.

Pipes, rigid or thin-wall conduit and the like may prevent installing plywood panels. The most annoying location for such obstructions is mid-height of the wall; this usually makes it impossible to slip a full-height shear panel behind the pipe. In this case a variation of the reverse-block method is handy, where you pre-attach a nailer to plywood segments that are then installed above and below the obstruction, with the nailers running parallel to it just above and below. Install a third plywood strip that connects to the nailers and you have a complete load-path. Figure 6-58 shows a case where the pipe turned a corner halfway along the wall, so this method was needed only along part of the wall.

Figure 6-58 *Left:* Pipe running at mid-height of wall obstructs plywood installation. *Center:* Finished installation; slot for pipe is boxed-out with flat 2x4 nailers pre-attached from the back of the panel before installation, then covered with a plywood strip; *Right:* View of the completed installation from the back of the wall. [An alternative connection (not applicable here because of the electrical cable in the way) would be to install LVL or kiln-dried blocking behind the pipe and have a horizontal plywood joint along the pipe with plywood edge-nailing to the blocks.]

If pipes or conduit run along the top plate of the wall you can usually coax them an additional ½" away from the framing to install plywood. For multiple pipes that obstruct the top plate you will likely need a customized solution. (Devising a connection from the floor diaphragm above that provides a complete load path would be a good question for a retrofit contractor certification exam.)

6.4.4 Partition walls that butt into shear walls

Non-bearing walls dividing off storage areas or utility closets often get in the way of where you want a shear wall. Usually it's easiest simply to remove the end of the partition wall, install the shear panels as needed, and replace the partition.

In some cases you may need to leave the intersecting wall framing in place. You can transfer shear forces through the "channel stud" of the intersecting wall as shown in Figure 7-39.

6.4.5 Diagonal braces from studs to floor joists

Some previous retrofits used diagonal braces installed as shown in Figure 6-59. While an engineer could design braces like these to adequately resist an earthquake, it would be much easier to install plywood bracing around the perimeter walls of the crawlspace. Furthermore, the braces shown will resist earthquake forces acting only in the direction of the joists. To resist forces perpendicular to the joists, we have to install plywood to the walls supporting the joists—which are now made quite inaccessible by the diagonal braces.

Figure 6-59 Diagonal braces are not considered best practice; these braces are in the way of installing plywood panels. Photo by Paul Barraza, GGASHI

Rather than cut shear panels to fit around the diagonal braces (and install new framing to provide for nailing at all panel edges) it is usually easiest to simply remove the diagonal braces. [It is possible that the braces were installed to address sagging or bouncing floor framing. If this is the case, one would expect the braces to support the joists close to the original cross-bridging, and for braces to occur at every other joist or even every joist. Supporting joists using diagonal braces is rather ineffective when compared to installing a new support beam, and also problematic for other reasons beyond our scope of discussion. If you suspect that braces are intended for more than earthquake resistance you should not remove them without consulting an engineer.]

6.4.6 "Sway braces" or "knee braces"

Some "soft story" construction has diagonal bracing at side walls intended to address the open front of the building; see Figure 6-60. Such bracing is rarely adequate to resist earthquake forces. An engineered "distributed knee-brace" system is under study at Cal Poly, but is much more complex and robust than these. If you can brace the perimeter walls with plywood, you can almost always permanently remove knee-braces with no strength loss. As discussed above, though, consult with an engineer before removing braces.

Figure 6-60 Knee braces—exposed (top) and covered with gypsum board (bottom)

6.4.7 Shear-transfer diaphragms

Shear transfer diaphragms are used where the load in a structure must be resisted by a member that is offset horizontally from the load. Where a floor extends past the braced wall below, the floor extension must act as shear transfer diaphragm. Figure 6-61 shows a sun-room supported on posts; the unbraced posts cannot resist horizontal earthquake forces, so all such forces must transfer through the sunroom floor diaphragm to the front wall of the house.

Think of a shear transfer diaphragm as a shear wall laid flat. As with any "shear element," the shear transfer diaphragm must have a force applied along each edge. When we construct a diaphragm out of plywood we need framing or blocking supporting each edge, and the plywood must be nailed along each edge. Figure 6-62 shows a couple of examples of shear transfer diaphragms; note that in most cases the diaphragm has to be only about one joist bay in width. In both cases shown, the plywood strip was first nailed to the top of a 2x8; this assembly was then installed in place and the plywood strip was nailed up to the bottom of the joist and blocks; lastly the 2x8 was connected to the foundation or shear wall, as the case may be. The detail in Figure 7-24 gives an example of specific shear transfer diaphragm construction requirements.

One common shortcoming of shear transfer diaphragms is the lack of blocking at the ends. Often, thicker plywood was used for the diaphragm, perhaps to account for the unsupported edges. Best practice would be to add full-depth joist blocking at both ends of the diaphragm in such cases, glued to the underside of the original subfloor, and nail the plywood up into the blocks.

6.5 Repair Methods for poorly-installed prior retrofit work

Engineers, contractors and home inspectors often come across retrofit work that is really close, but does not quite reach the goal of being effective. In many cases there are easy—or at least reasonable—ways to "retrofit the retrofit" so it will work as originally intended. This section presents

Figure 6-61 The floor of this sunroom supported on porch posts must transfer lateral loads back to the main house wall.

Figure 6-62 Shear transfer diaphragms in earthquake retrofits; Top: from joist to foundation; Bottom: from joist to shear wall.

methods for some of the most common problems with prior retrofit work.

6.5.1 Lack of blocking behind plywood panel joints

Shear forces have to transfer from one plywood panel to any adjacent panel. Generally this means nailing the panel edges to a common framing member, either a stud or a row of blocks. Acres of plywood have been installed with no framing behind horizontal (and sometimes vertical) panel joints. If adjacent panels are not nailed to a common framing member you have two options: Install a new framing member behind the joint and nail the panels to it, or bridge across the joint another way—both are described next.

6.5.1.1 Install retrofit framing behind panel joints

Most often, horizontal joints are the ones that need retrofit blocking. When the back of the wall is accessible you can simply install new blocks. As described earlier for joist blocking, use kiln-dried or engineered lumber, cut to fit snugly between the studs, and end-nail or toe-nail all blocks. Installing blocks flat-wise gives a wider nailing surface for attaching the plywood edges. Since flat blocks will split more easily between the plywood panels, I recommend using 3x or thicker material. Nail the plywood to the new blocks with the same size nails as used for the original edge-nailing (or upgrade as necessary, discussed elsewhere) and at the typical edge nailing

Figure 6-63 Studs stitched together with structural screws after the original construction to transfer shear forces across a vertical panel joint. Plywood patch with "54" label is glued across an un-blocked horizontal panel joint. Access to the other side of this wall was blocked by kitchen cabinets.

If panels are not supported at their vertical edges, install a new stud or post behind the joint. In some cases you may need to install a sister stud against an existing stud and nail the plywood to the sister (see Figure 6-63). [NOTE: In most cases the original framing is already supporting vertical loads; we do not need another "stud" (full-height bearing member), but rather a doubled member that can provide a nailing surface for a plywood edge. In the case shown in Figure 6-63, adding a stud would have been impossible because of the existing pipe; if a new nailing surface was needed, "scabs" above and below the pipe would provide adequate nailing surface.] Connect the two members together with 16-penny common nails at the same spacing as the plywood edge nailing (this is generally sufficient where plywood is attached with 10-penny nails; for 8-penny plywood nails you may use 16-penny sinkers). In the most annoying cases, you have to add sister studs on both sides of the existing studs, as when the edge-nail spacing requires a member at least 2½" thick but one was not provided.

Splitting wood members is always a concern. Pre-drill for nails as needed to prevent splitting, and stagger rows of nails. Don't use too many nails.

Using structural screws can help prevent splitting existing dry lumber because the screws have some "self-drilling" abilities. Structural screws are generally tested to much higher loads than nails. For Simpson's "SDS" or "SDWS" screws, sistering studs with about one screw for every four 8-penny nails (one screw per three 10-penny nails) roughly matches the capacity of the plywood nailng and the stud connection. As an example, if the plywood edge-nailing is 8-penny nails at 3" on center you would sister studs together with screws at 12" spacing.

For general explanations of sistering studs and joining studs at tie-downs, see Detail P1 (Figure 7-41).

Often the back of the wall is not accessible, and the plywood obstructs installing blocks. In this case you will need to use one of the following methods.

6.5.1.2 Bridge across the panel joint on the 'front' of the wall

Sheet metal "blocking:" *APA—the Engineered Wood Association* tested sheet-metal "blocking" attached to the face of plywood using structural staples as a method to transfer shear across the panel joint. More information is available in the APA's Technical Note #N370C. Figure 6-64 shows a case inside a finished home in which the horizontal plywood joint had no blocking behind it, and access was not available from the far side of the wall.

Figure 6-64 Sheet metal attached to plywood with structural staples bridges across an un-blocked horizontal panel joint. Ideally the staples would not dimple the sheet metal. (A tiled bathroom on the opposite side of the wall prevented access to install blocking behind the plywood joint.)

Glued plywood blocking: Plywood strips can be glued across the joint as another option. Minimum width of strips should be as follows: 6" to substitute for 6-inch edge nailing; 9" for 4-inch edge nailing; 12" for 3-inch EN; and 18" for 2-inch EN. Apply continuous ¼" beads of construction adhesive spaced no more than 2 inches apart under the plywood strips. Secure plywood in place with nails and/or self-drilling screws spaced at 8" along both edges. Figure 6-63 shows a plywood patch glued across a horizontal panel joint. This retrofit project was a rare case in the living area of a house, where it was important to minimize disturbance to the far side of the wall.

6.5.2 Backing not provided for nailing bottom edge of plywood

Many a retrofit installer was confounded by the mudsill being wider than the cripple studs. Mudsill blocks were often not installed. Usually the plywood just ends at the top of the mudsill with no connection at all; sometimes the installer toe-nailed the bottom edge of the plywood to the mudsill, as shown in Figure 6-65. In either case, the bottom edge of the panels must be properly secured to the mudsill.

If the back of the wall is accessible, you can install mudsill blocks and nail the plywood to them. Even if you can access the rear of the wall it may be easier to cut the mudsill flush with the face of the plywood, and then bridge the joint as discussed in the previous section. NOTE: Do not attach plywood shear panels to the mudsill with adhesive.

Figure 6-65 BAD: Plywood installed with no connection along bottom edge except random toenails.
Photo by Matt Cantor, GGASHI

REPAIRED: Plywood strip with reverse-block at bottom installed to previous plywood with structural staples.
(Paint covers the staples; they hide in the four rows of small dimpled slots.)

6.5.3 Mudsill anchors not installed at shear wall sections

Depending on the era of construction, size and configuration of the house, omission of mudsill anchors may not be an enormous problem (see Section 4.1.2.4). However, without "x-ray vision" it is impossible to know this. I always recommend installing more anchors if you can afford it, especially if there is much doubt about existing connections.

As discussed in Section 6.2.1.1 you can connect the mudsill to the footing outside of the length of plywood bracing panels using suitable methods. You can also connect through the plywood with connectors such as the UFP10, though this is not the optimum connection (see Section 6.2.1.9.1)

6.5.4 "T1-11" siding nailed along only one edge

Plywood siding was commonly installed by attaching only the overlapping edge on four-foot wide panels. This installation method does not transfer shear forces between adjacent panels.- The near-collapse shown in Figure 6-66 was blamed on improper nailing at siding joints.

Using structural screws can help prevent splitting existing dry lumber because the screws have some "self-drilling" abilities. Structural screws are generally tested to much higher loads than nails. For Simpson's "SDS" or "SDWS" screws, sistering studs with about one screw for every four 8-penny nails (one screw per three 10-penny nails) roughly matches the capacity of the plywood nailng and the stud connection. As an example, if the plywood edge-nailing is 8-penny nails at 3" on center you would sister studs together with screws at 12" spacing.

For general explanations of sistering studs and joining studs at tie-downs, see Detail P1 (Figure 7-41).

Often the back of the wall is not accessible, and the plywood obstructs installing blocks. In this case you will need to use one of the following methods.

6.5.1.2 Bridge across the panel joint on the 'front' of the wall

Sheet metal "blocking:" *APA—the Engineered Wood Association* tested sheet-metal "blocking" attached to the face of plywood using structural staples as a method to transfer shear across the panel joint. More information is available in the APA's Technical Note #N370C. Figure 6-64 shows a case inside a finished home in which the horizontal plywood joint had no blocking behind it, and access was not available from the far side of the wall.

Glued plywood blocking: Plywood strips can be glued across the joint as another option. Minimum width of strips should be as follows: 6" to substitute for 6-inch edge nailing; 9" for 4-inch edge nailing; 12" for 3-inch EN; and 18" for 2-inch EN. Apply continuous ¼" beads of construction adhesive spaced no more than 2 inches apart under the plywood strips. Secure plywood in place with nails and/or self-drilling screws spaced at 8" along both edges. Figure 6-63 shows a plywood patch glued across a horizontal panel joint. This retrofit project was a rare case in the living area of a house, where it was important to minimize disturbance to the far side of the wall.

Figure 6-64 Sheet metal attached to plywood with structural staples bridges across an un-blocked horizontal panel joint. Ideally the staples would not dimple the sheet metal. (A tiled bathroom on the opposite side of the wall prevented access to install blocking behind the plywood joint.)

6.5.2 Backing not provided for nailing bottom edge of plywood

Many a retrofit installer was confounded by the mudsill being wider than the cripple studs. Mudsill blocks were often not installed. Usually the plywood just ends at the top of the mudsill with no connection at all; sometimes the installer toe-nailed the bottom edge of the plywood to the mudsill, as shown in Figure 6-65. In either case, the bottom edge of the panels must be properly secured to the mudsill.

If the back of the wall is accessible, you can install mudsill blocks and nail the plywood to them. Even if you can access the rear of the wall it may be easier to cut the mudsill flush with the face of the plywood, and then bridge the joint as discussed in the previous section. NOTE: Do not attach plywood shear panels to the mudsill with adhesive.

Figure 6-65 BAD: Plywood installed with no connection along bottom edge except random toenails.
Photo by Matt Cantor, GGASHI

REPAIRED: Plywood strip with reverse-block at bottom installed to previous plywood with structural staples.
(Paint covers the staples; they hide in the four rows of small dimpled slots.)

6.5.3 Mudsill anchors not installed at shear wall sections

Depending on the era of construction, size and configuration of the house, omission of mudsill anchors may not be an enormous problem (see Section 4.1.2.4). However, without "x-ray vision" it is impossible to know this. I always recommend installing more anchors if you can afford it, especially if there is much doubt about existing connections.

As discussed in Section 6.2.1.1 you can connect the mudsill to the footing outside of the length of plywood bracing panels using suitable methods. You can also connect through the plywood with connectors such as the UFP10, though this is not the optimum connection (see Section 6.2.1.9.1)

6.5.4 "T1-11" siding nailed along only one edge

Plywood siding was commonly installed by attaching only the overlapping edge on four-foot wide panels. This installation method does not transfer shear forces between adjacent panels.- The near-collapse shown in Figure 6-66 was blamed on improper nailing at siding joints.

Figure 6-66 Grooved plywood siding (one style name is "T1-11") that failed under earthquake loading because it was not nailed properly. Plywood must be nailed along each edge of each panel. FEMA Photo

BAD: Close-up of grooved plywood siding nailed on only one side of the lap-joint. Most siding is installed in in this manner.

Siding nailed both sides of joint; even in the best case, the thin overlap does not provide secure nailing.

"But we've always done it this way." Grooved plywood siding was introduced 40 or 50 years ago. This "two-in-one" product saved builders from installing a layer of structural sheathing and then a layer of finished siding. They quickly figured out that they could save even more time by nailing only the over-lapping panel edge at vertical joints. APA—The Engineered Wood Association does not recommend this practice, as it does not transfer shear forces across the panel joint. Yet a great majority—in my experience about 80 percent—of grooved plywood siding is installed with only one edge nailed, even in earthquake-prone California where every builder should know better.

Figure 7-38 shows how to correct deficient panel edge nailing. The most important area to correct deficient plywood siding nailing is usually below the first floor. Since adding more panel edge nailing is relatively easy, I strongly recommend increasing the nailing on both sides of the panel joints.

Improving the nailing of siding is pointless if the siding has decayed. The overlapping segments of plywood siding are relatively thin and prone to fungal attack if not protected from water. Grooved plywood siding was intended to save the builder money. When projects are driven by the desire to keep expenses down, good construction practices besides nailing may have been sacrificed as well—such as careful painting.

6.5.5 Foundation caps without connection to original footings

Current building codes require eight inches of vertical separation between the ground and any wood that is not decay-resistant. Many houses built before this requirement was in place had "foundation caps" installed. The cap is a layer of concrete poured on top of the original footings or stemwalls, varying from a few inches to a foot or more tall. Figure 6-67 shows a typical foundation cap; the original footing appears to be only about four inches tall, with perhaps very little extension below ground level. Quality of work for foundation caps varies tremendously. In some cases the cap was installed without even bothering to remove the original mudsill. If the original foundation is deteriorating severely, it's probably time to construct new footings; sometimes the foundation should have been replaced rather than capped.

Unless you can verify that the cap was "pinned" to the original footing (using steel rods set in holes in the original footings) you should not depend on the cap to stay in one place during an earthquake. You do not want to secure your house to the cap only to have the cap slide off the original foundation.

Figure 6-68 shows a foundation cap secured to the original footing with a section of 5/8" diameter steel reinforcing bar set with epoxy. Such bars could be viewed somewhat as mudsill anchor rods, except that the strength of a connection from concrete to concrete has a much higher capacity than concrete to wood mudsill. Connection pins in the foundation cap could safely be installed at twice the mudsill anchor spacing.

Figure 6-67 **Usually Bad**: Capped footing—the top two-thirds of this footing was installed decades after the original construction. Foundation "caps" are rarely connected well to the original foundation.

Figure 6-68 Fixed: Retrofit connection for a capped footing; steel pin set with epoxy into holes drilled through the cap and into the original footing.

uplift/downward forces that the steel frames can impose. Installers must follow all of the design engineer's requirements plus instructions from the manufacturer to achieve the intended strength of the frame.

6.6.2.2 "Fast Track" steel foundation system

The *Fast-Ttrack* system (see http://www.anchorpanel.com/) was originally developed to support mobile homes (called "manufactured houses" if you are trying to sell them) around their perimeters. The system uses 16-gauge sheet metal with deeply corrugated ribs to take the place of a cripple wall and sheathing, as shown in Figure 6-73.

The corrugated panels are pre-cut by the manufacturer to the height needed; the bottom is bent to key into the footing. For retrofit installations a house can be shored up while new footing trenches are excavated. The house shown was originally built on a post-and-pier system (no perimeter foundation or cripple walls). The contractor raised and re-leveled the house; then he removed the old posts and piers and dug trenches for new footings.

Once the footing trenches were excavated, the Fast Track system was installed. First a 16-gauge steel track was fastened to the underside of the floor framing system. Next the corrugated panels were attached to the track with structural sheet metal screws (Figure 6-74), suspended several inches above the bottom of the trench. Panel seams were sealed and stitched with screws (Figure 6-75). Then the concrete was placed for the footings (Figure 6-76. The Fast Track panels were painted and now serve as the skirting around the house.

Testing established very high shear capacities for Fast Track panels when compared to plywood, even when combined with gravity loads greater than typical found in houses.
As with any proprietary product, though, current code listing reports may be required before your particular jurisdiction will allow using this system.

Figure 6-73   Completed "Fast Track" installation under existing house.

Figure 6-74   Interior view of track connection at wood framing (screws into framing not visible).

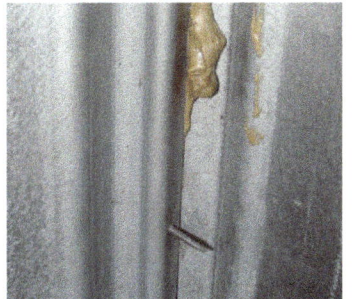
Figure 6-75   Connection between corrugated panels with sealant.

Figure 6-76   Base of steel panel embedded in concrete footing.

Chapter 7 Construction details

"Realistic details that are simple and are easily accomplished by tradespeople with a wide range of skill levels, will be the most successful in achieving the structural integrity goals set forth in the Preface to the [building code]." --Rick Tanis, American Plywood Association, May 3, 1990 memorandum to the State of Washington Association of Building Officials regarding lateral strength of buildings. In many retrofit situations "simple" connection details will just not do the job. Therefore only higher-skilled workers should attempt to install them.

The details included on the following pages have almost all been used in actual construction projects. They may not apply to YOUR construction project. These are presented only as examples; only competent professionals should determine how and whether to apply any of the following to a specific project, or modify them for specific use.

Comments are included with some of the details. The detail designations such as "W1" are for reference and do not follow any noticeable order.

In some cases photos are included to illustrate where the connection is used.

7.1 Mudsill Connections to Footings

7.1.1 Footings with cripple walls

Detail W1—Flush-cut mudsill

This detail will work equally well for blocks between the joists or joists butted into a "rim joist".

Stitch-nailing between the top plates helps them act as a single member. Plywood edge nailing can then be staggered between the two top plates—see Detail P1 for important notes about nailing.

Tie-downs *may* be required; if so, they should be indicated on the foundation plan.

Locate anchor bolt as indicated to provide more edge distance in the concrete footing. Use a slotted plate washer and install with edge flush with cut edge of mudsill.

Keep plywood slightly above original concrete foundation so that it will not wick moisture, and so that when the cripple wall racks back and forth the plywood can rotate slightly before contacting the concrete.

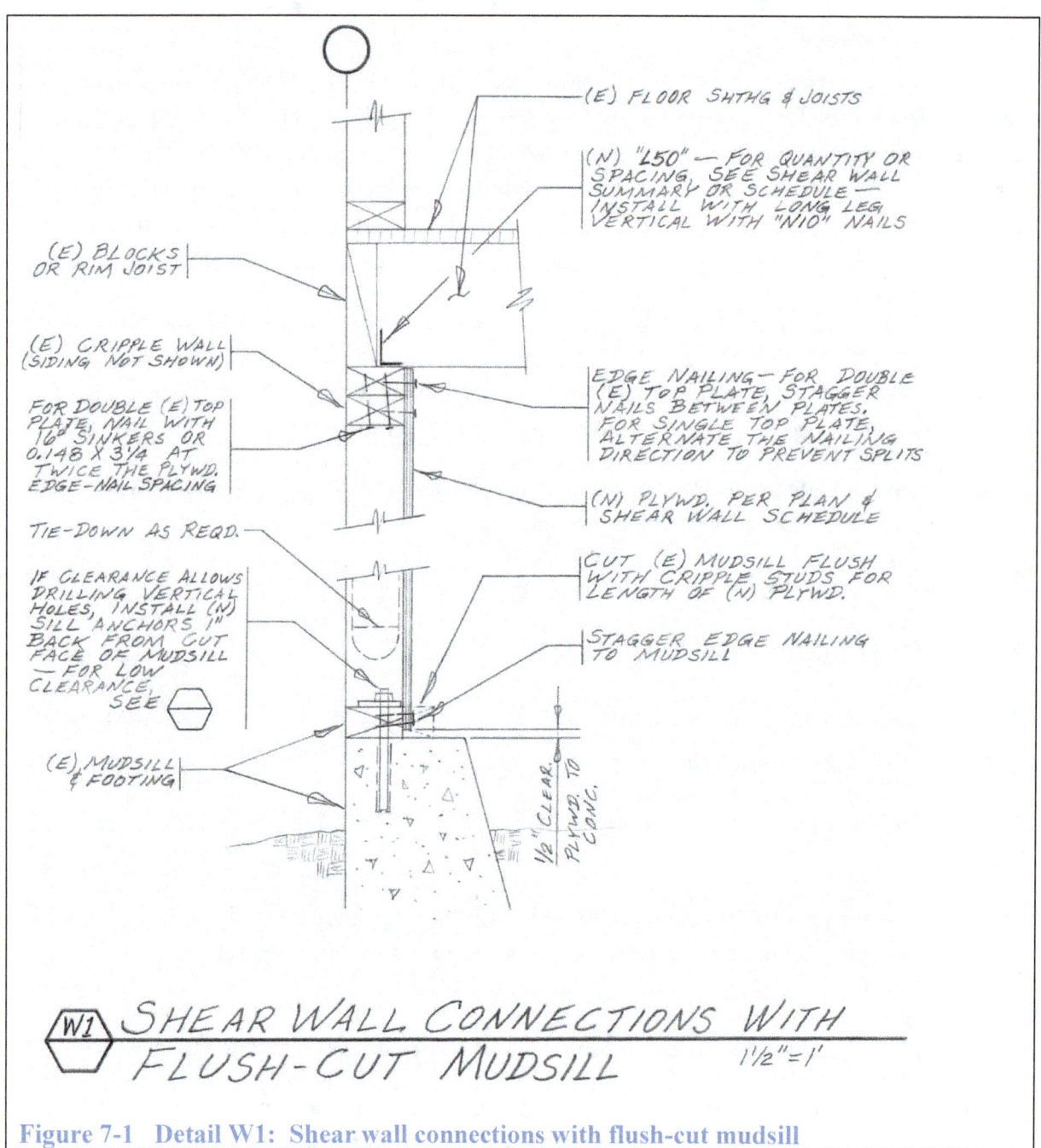

Figure 7-1   Detail W1: Shear wall connections with flush-cut mudsill

Detail C1—Typical Shear Wall Connections to Footing

Use when mudsill is not more than ¼" wider than studs.

If studs are not flush with mudsill, install new studs at all vertical panel edges that are flush with mudsill

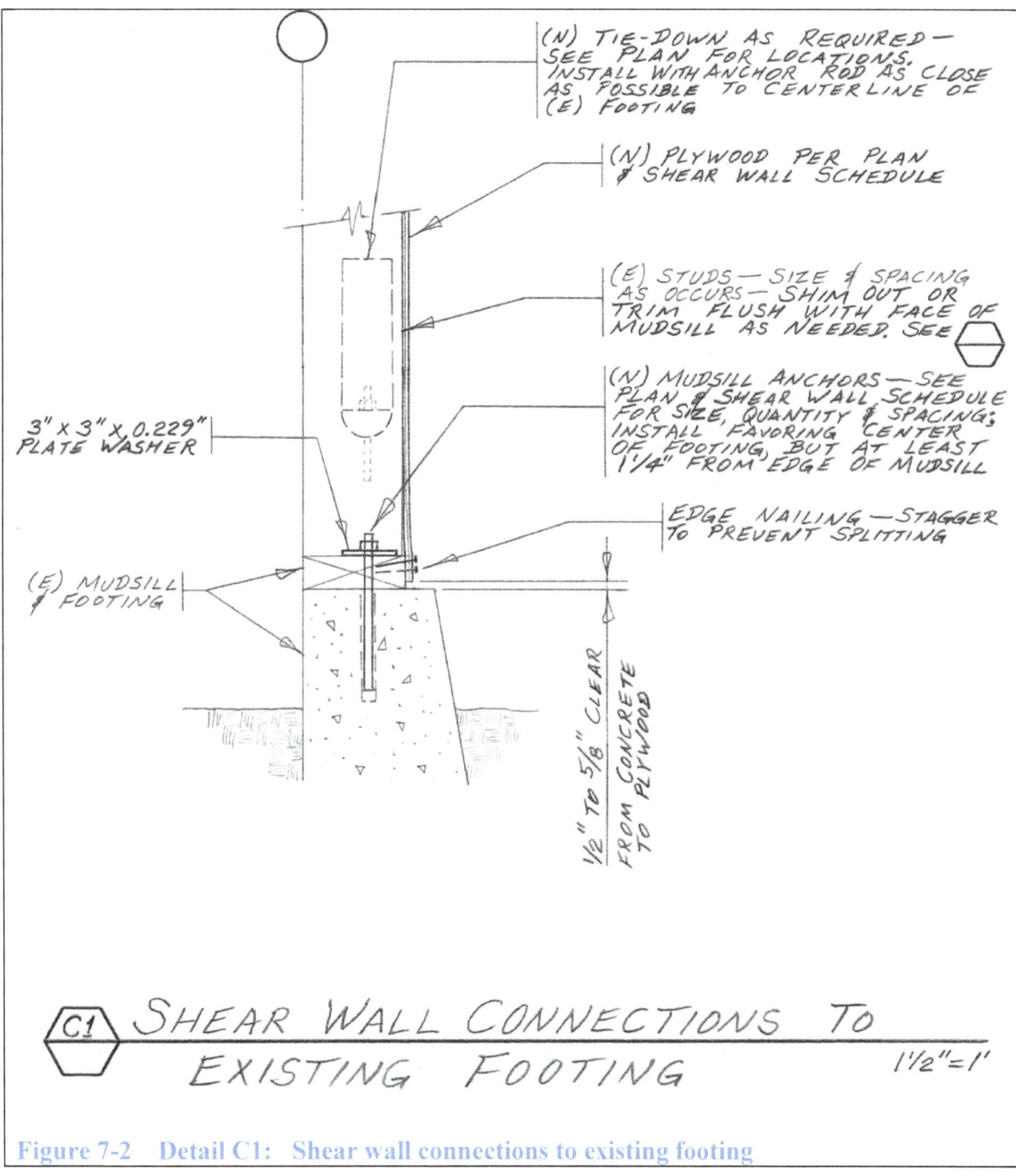

Figure 7-2 Detail C1: Shear wall connections to existing footing

Detail C2—Mudsill Blocking Connections

Flush-cut mudsill method is preferred. If you must use the mudsill block method, use screws (not nails) to connect the blocks. Use 3x or 4x blocks to reduce chances of splitting.

Blocks do not need to be more than 4" wide; 3x4 blocks may fit more easily into 2x6 stud spaces against irregular siding or stucco.

Retrofit anchors may be installed through blocks; place screws to allow room for plate washer installation.

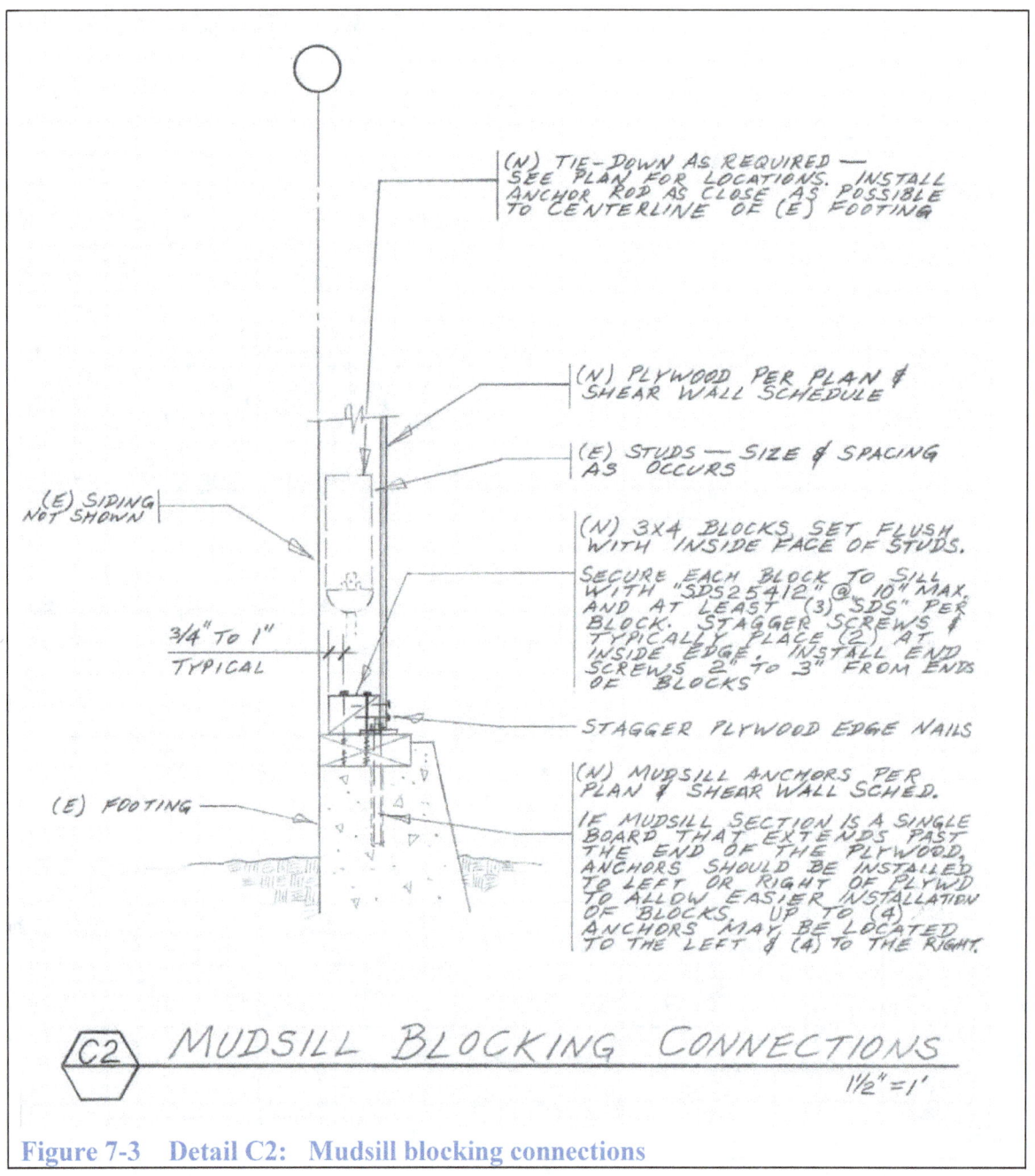

Figure 7-3   Detail C2:   Mudsill blocking connections

Detail C4—Mudsill Connection with KC Metals RFA Anchor

Use when footing is too wide for proper installation of UFP10.

Plywood bracing (shear) panels cannot be installed at same location as RFAs—use mudsill as a collector to transfer loads from the shear panels to adjacent foundation segments; strap joints in mudsill as needed.

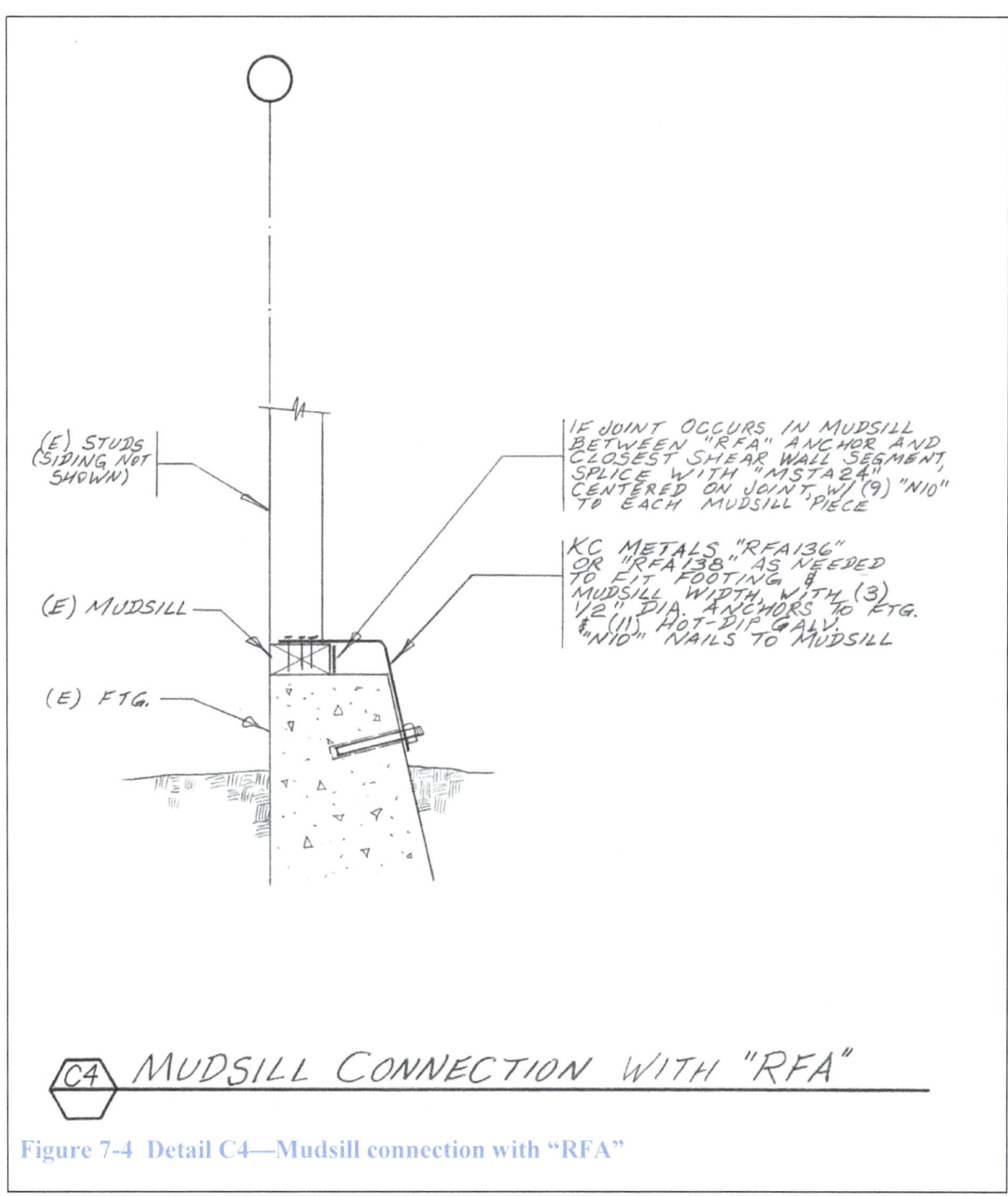

Figure 7-4 Detail C4—Mudsill connection with "RFA"

W2—Reverse Block Method

Use this detail if the mudsill is set too far back from the face of the footing to use a UFP10 attached directly to the original mudsill.

If the joist bay is longer than 6' with no existing blocks or cross-bridging, install blocks between the end joist and the second joist back as shown in Detail F2.

Connectors from Reverse Block to new mudsill must be designed to match the capacity of the shear panels.

See notes for Detail W1 for more information. Blocking at last joist bay keeps the end joist from rolling over, which can occur when several thousand pounds of force are applied along the joist and the whole house is shaking violently.

Detail W3—Shimmed Top Plate & End Studs

Use this detail if the total plywood height will be less than 24." Panel edge nailing is needed along only the top and bottom edges and at the panel ends. (This is the same as a 24" wide shear wall, only rotated 90 degrees.)

Shims may be increased to any thickness needed—but do not use more than two layers of material, and glue & nail them together.

Mudsill anchors will likely be UFP10s or RFA136s, installed outside of the length of plywood.

Detail W2—Reverse-Block Shear Wall at Extra-wide Existing Footing

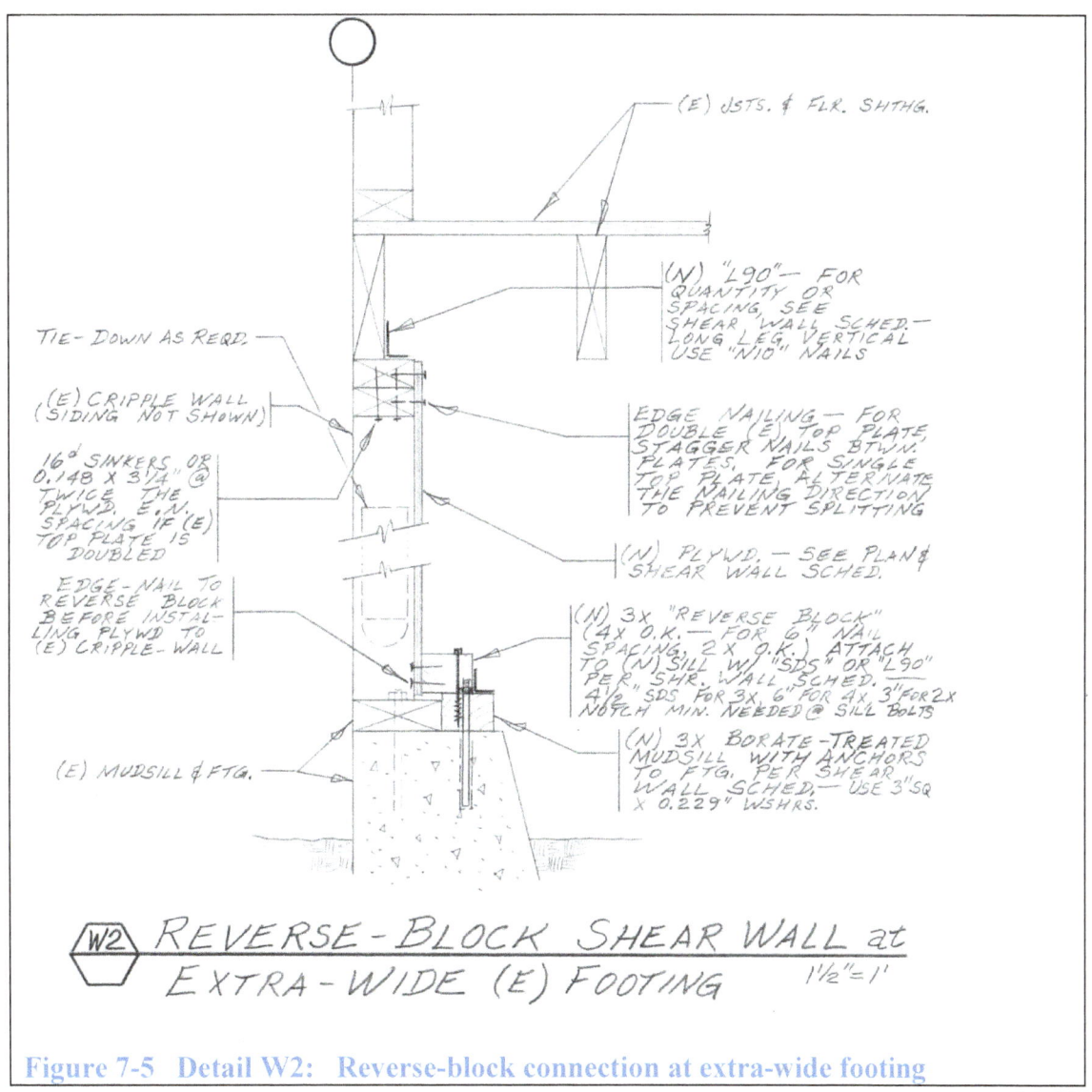

Figure 7-5 Detail W2: Reverse-block connection at extra-wide footing

Detail W3—Top Plates Shimmed out to Match Width of Mudsill

At all vertical panel edges, provide new studs or posts with their edges flush with the mudsill and shimmed top plate.

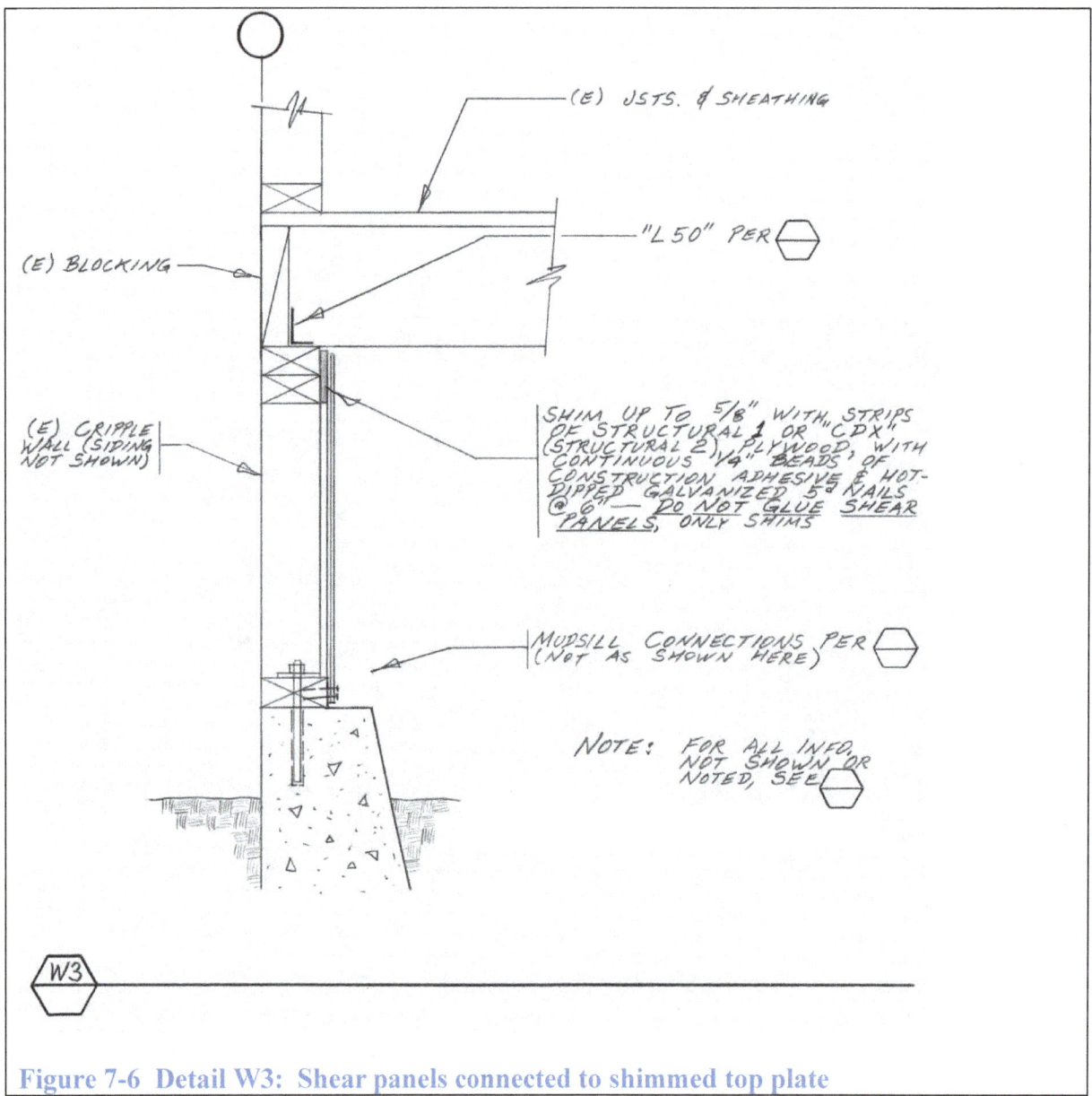

Figure 7-6 Detail W3: Shear panels connected to shimmed top plate

Detail W5—New Cripple Wall and Footing

Use when existing wall length or location cannot provide sufficient strength. Align under an existing joist so you can nail the new shear plywood directly to the face of the joist.

Typically connect the new footing to existing footings if possible, with rebar doweled into the existing concrete.

Provide tie-downs as needed (they will be much more effective when connected to a new, reinforced footing than if installed on an old footing).

Joist blocking at the top of the wall helps keep the end joist from rolling over, which can occur when several thousand pounds of force are applied along the joist and the whole house is shaking severely.

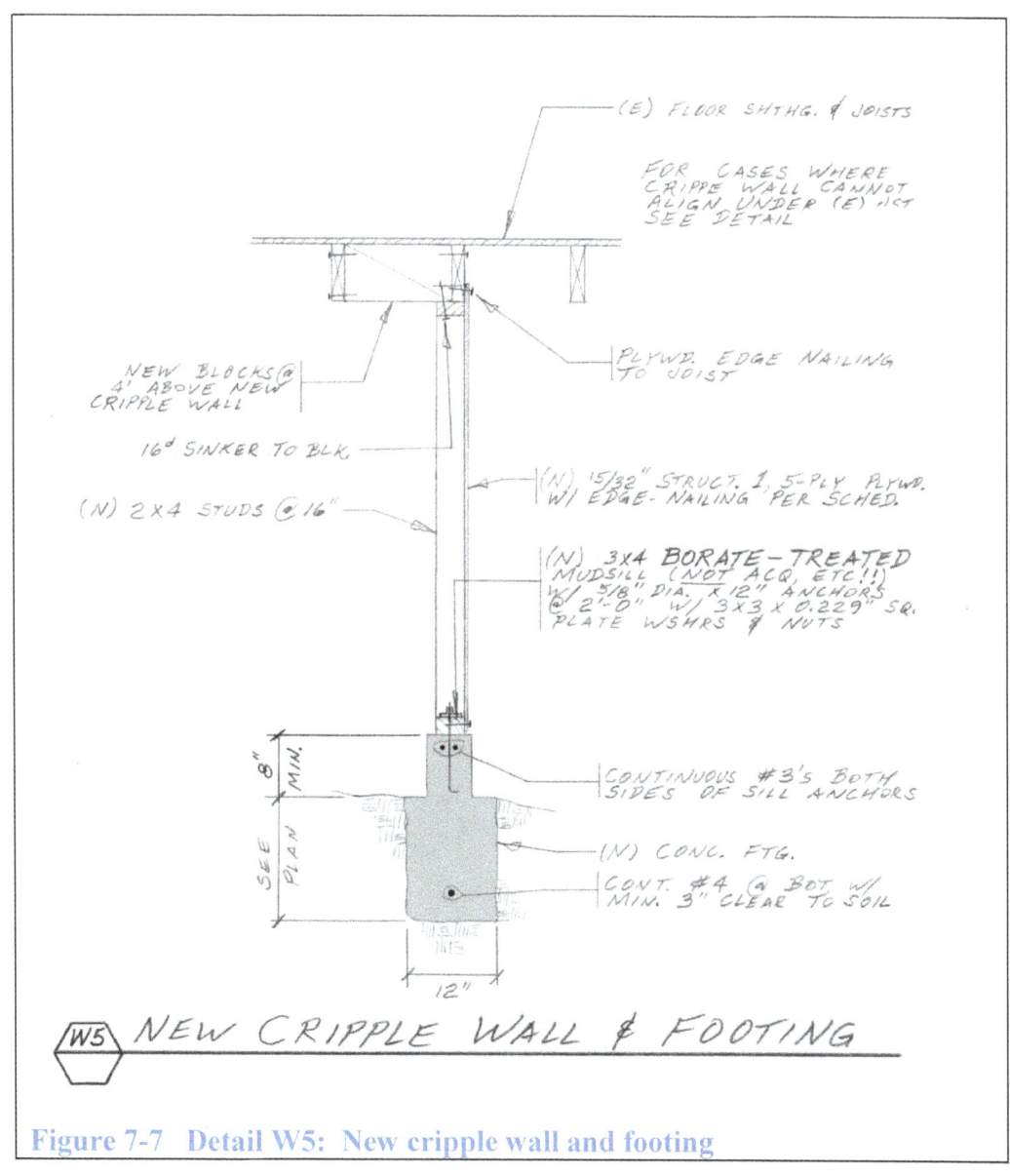

Figure 7-7   Detail W5: New cripple wall and footing

Balloon Framing Connections

These connections use "Reverse Blocks" and plywood blocking. Anchorage to mudsill may vary to suit actual conditions.

Shear panels should not extend below mudsill unless they are isolated from the concrete.

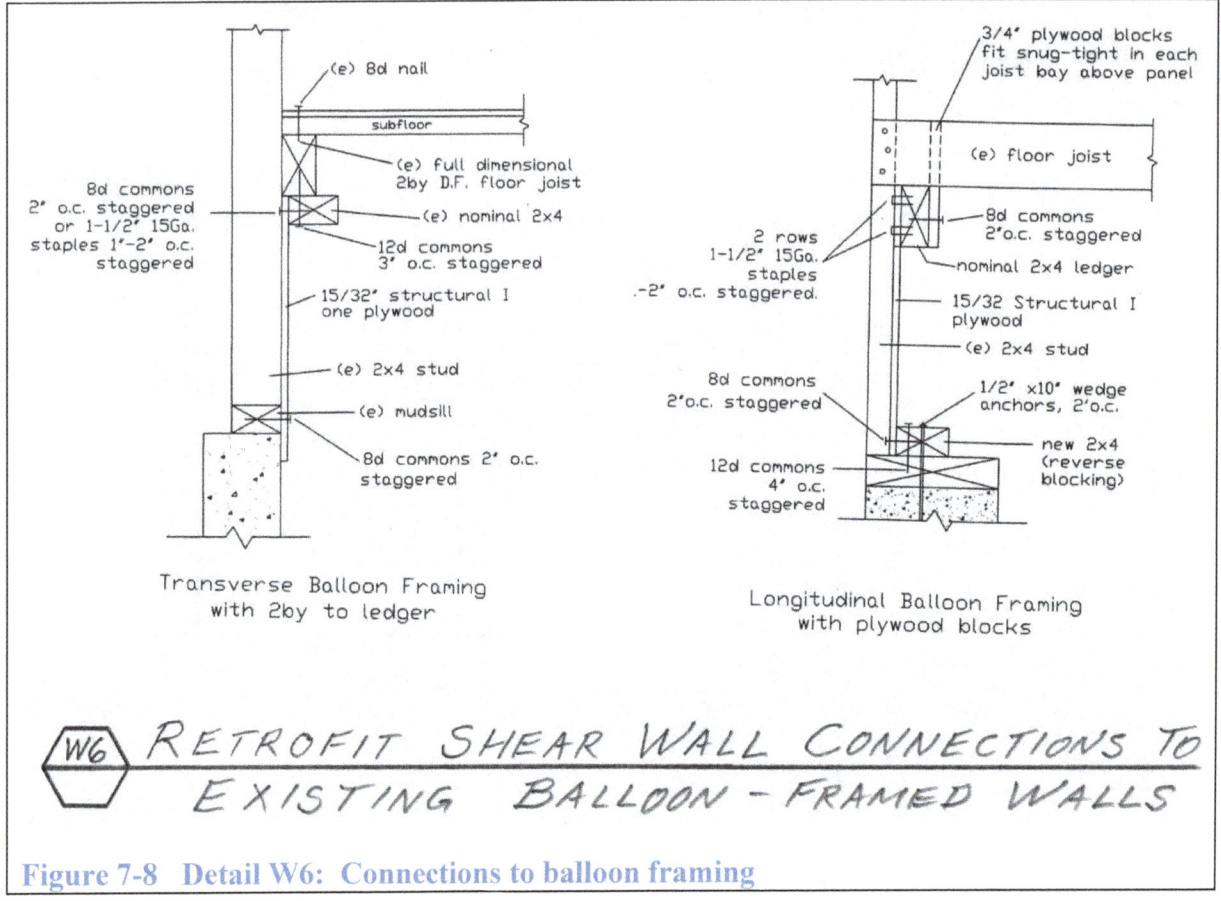

Figure 7-8   Detail W6: Connections to balloon framing

7.1.2 Footings without cripple walls

The following details apply to houses with framing supported directly on the mudsills.

Detail B1—UFP10 at Perpendicular Joists

Use where mudsill is inset no more than 2½" from the face of the concrete (or as otherwise required by foundation connector manufacturer).

For rim joist, use L90s or equivalent; for blocks less than 30" long, use L50 or equivalent.

Detail B2—UFP10 at Parallel Joists

Use where mudsill is inset no more than 2½" from the face of the concrete (or as otherwise required by foundation connector manufacturer).

Installing blocks every 4 to 6 feet would help stabilize the end joist from rolling over as the house is shaking around.

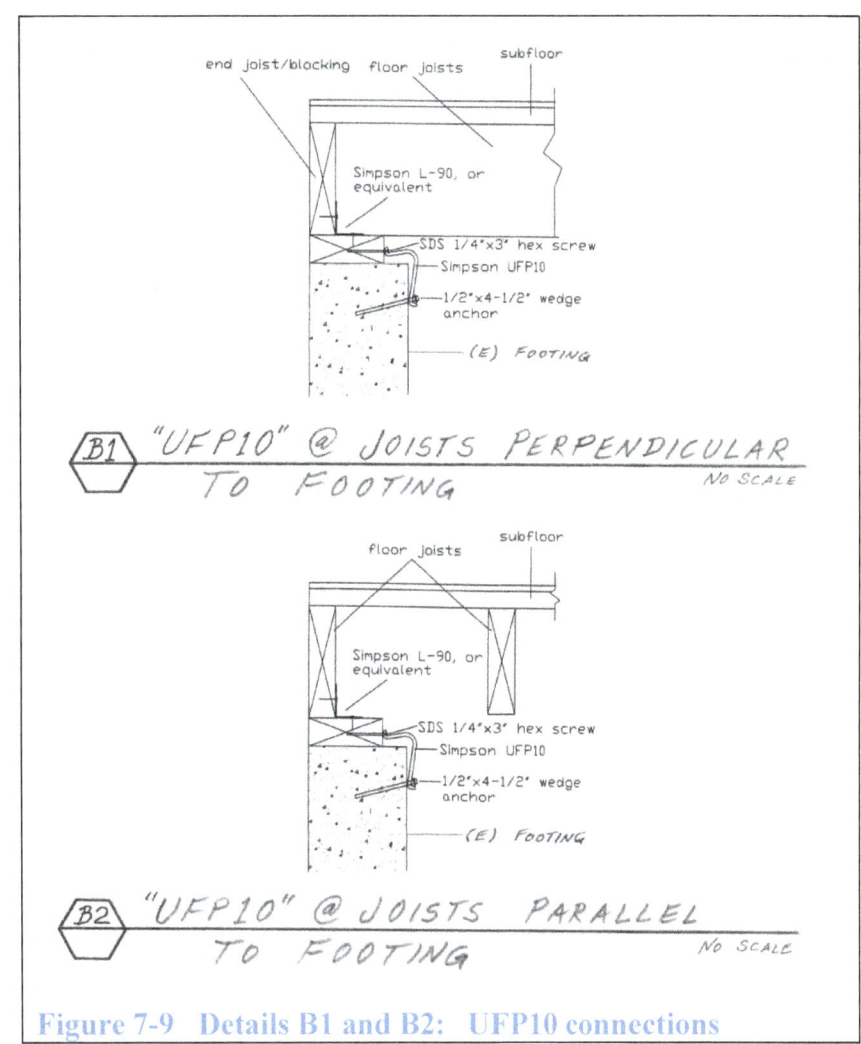

Figure 7-9 Details B1 and B2: UFP10 connections

Detail B3—Connection to Footing with Deeply Recessed Mudsill

Where the existing mudsill is "out of reach" of a UFP10 or similar foundation connector, install a new member to the face of the footing. Provide moisture barrier between concrete and wood.

Install new joist blocking with adhesive to floor sheathing and STTs to new side-bolted foundation member.

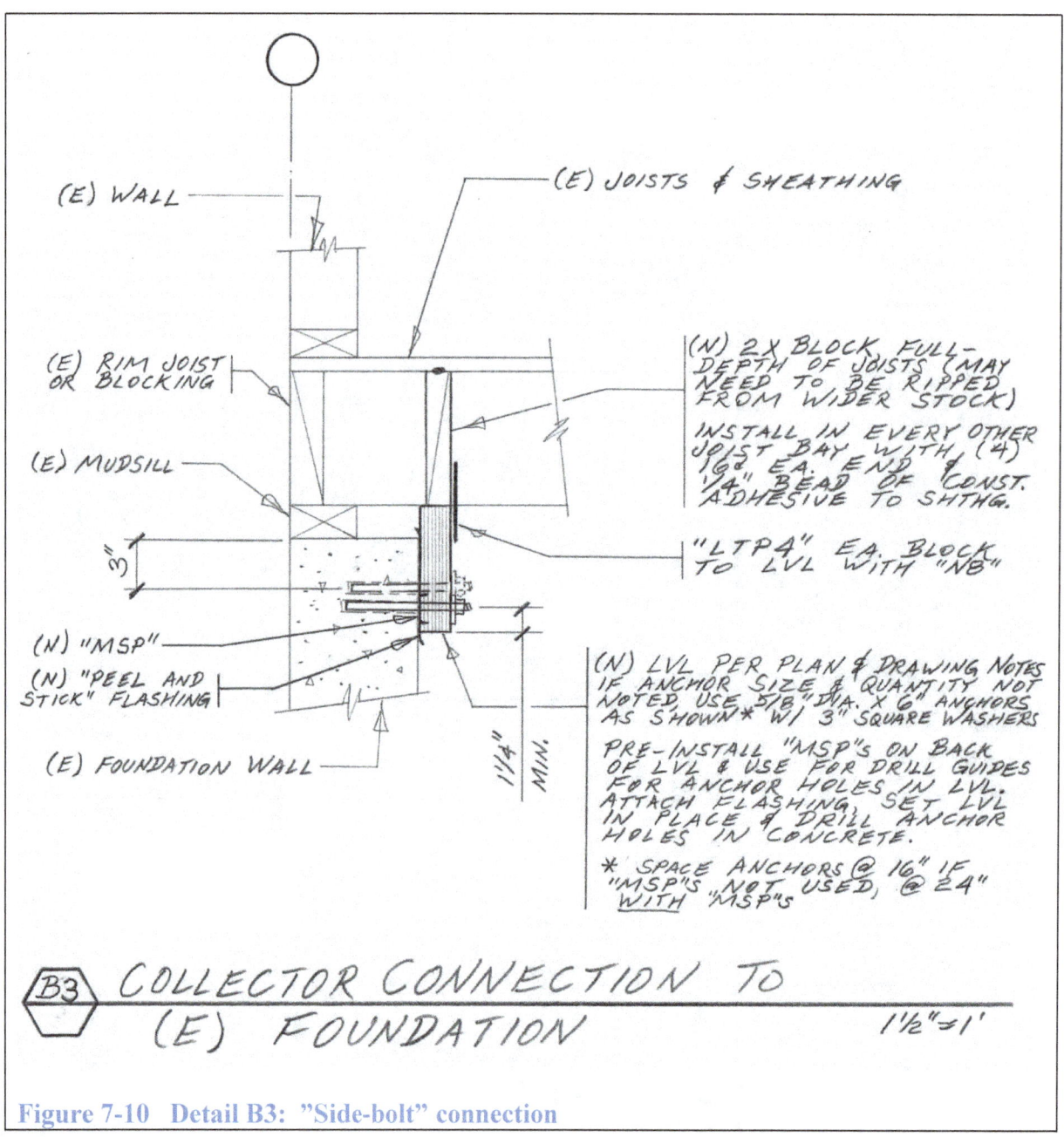

Figure 7-10   Detail B3: "Side-bolt" connection

Detail B4—Hip-and-Shoulder Footing

Blocking is essential in order to keep joist from rolling over and splitting at the H10 connection. (NOTE: The H10 was discontinued around 2013 and replaced with the H10A)

UFP10s or other foundation connectors may be used if the lower mudsill is not flush with the face of the concrete.

If the blocks are installed far enough inside of the footing to provide access for a palm nailer, hammer, slide-hammer, etc, then nails may be used instead of screws to connect to the joists.

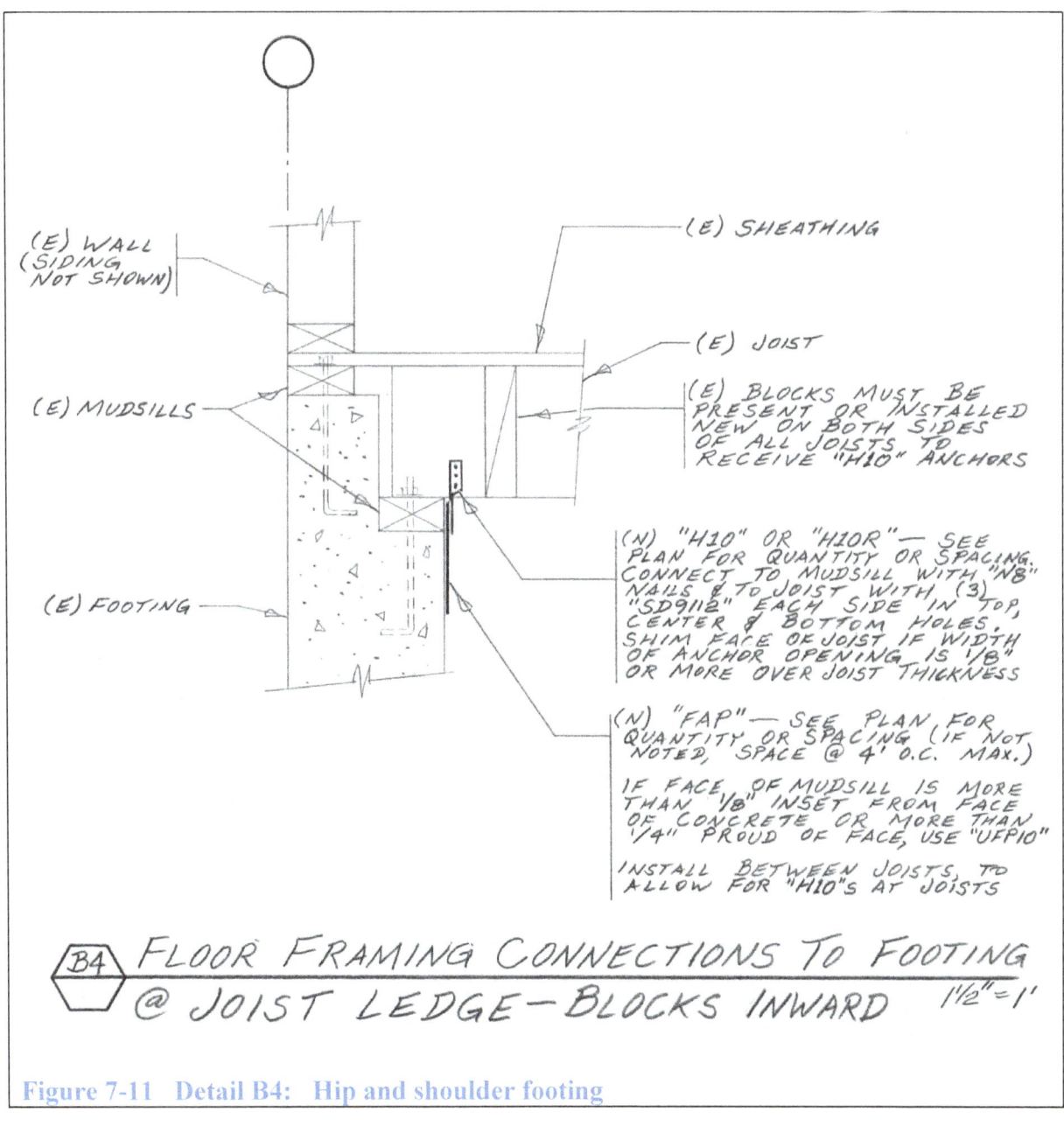

Figure 7-11 Detail B4: Hip and shoulder footing

Detail B5—Hip and Shoulder Footing

Use when blocks are installed flush with the mudsill. See notes for Detail B4 for additional information.

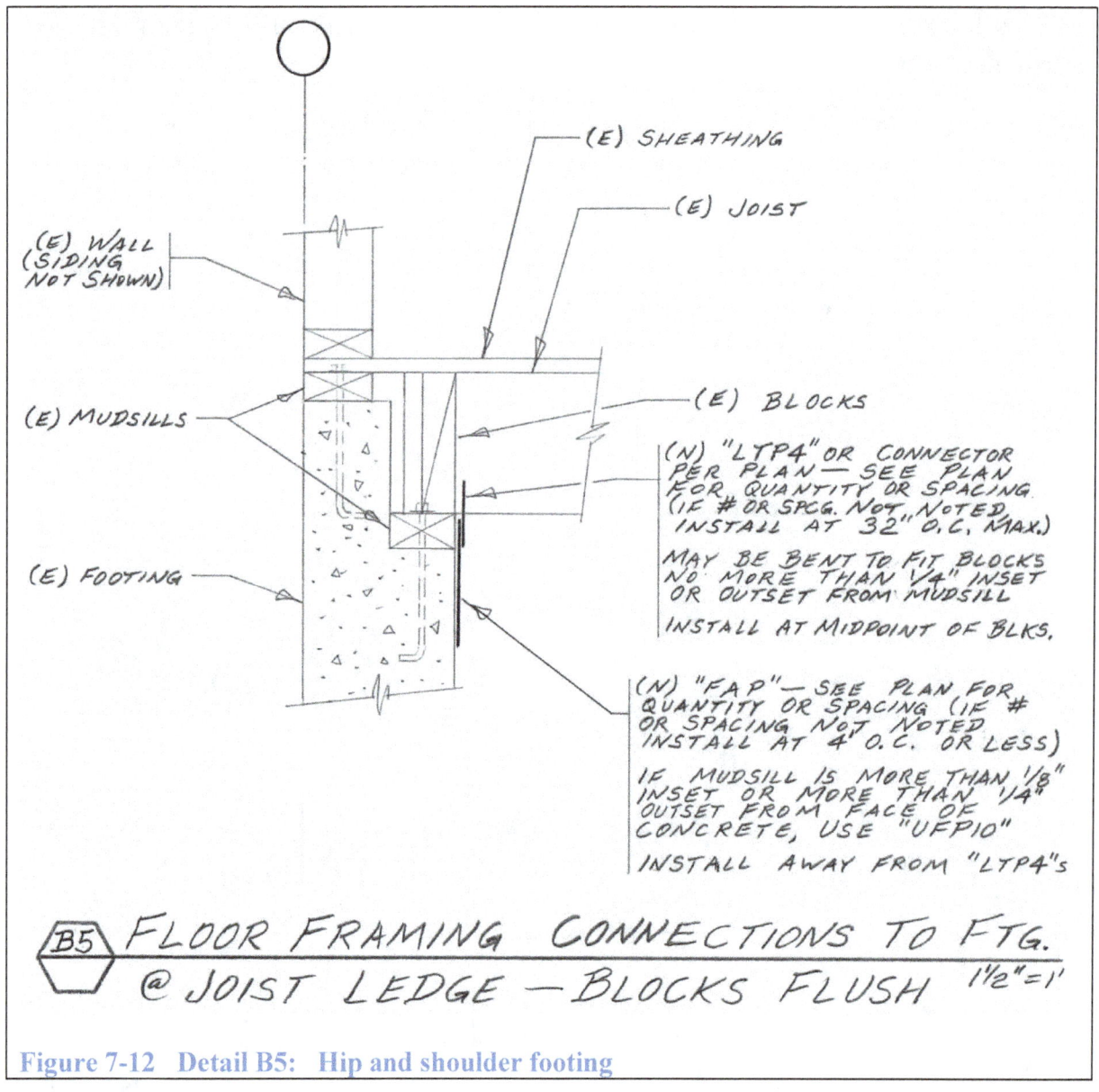

Figure 7-12   Detail B5:   Hip and shoulder footing

Detail B6—Connection at Multiple Mudsills

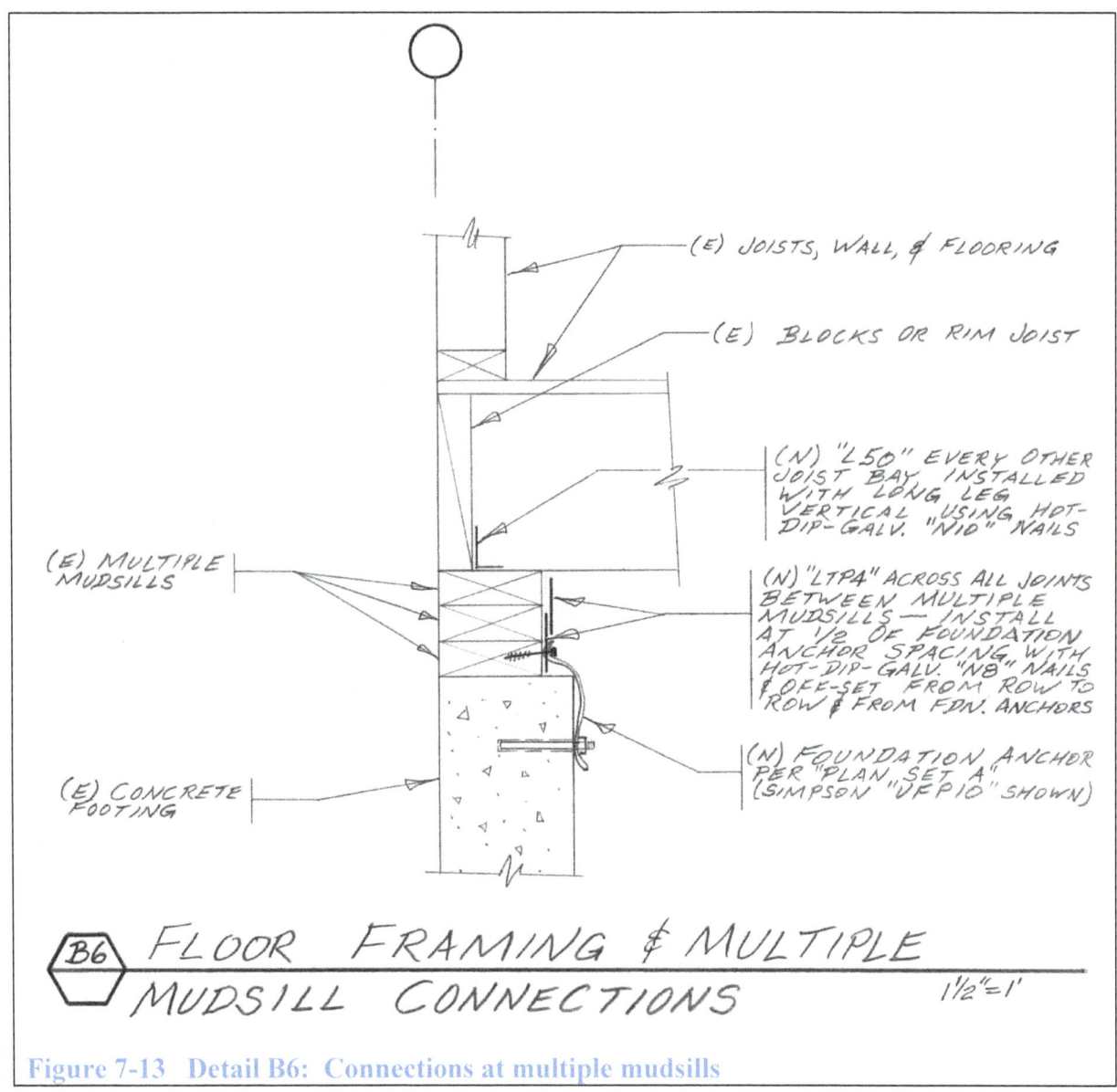

Figure 7-13 Detail B6: Connections at multiple mudsills

7.2 *Cripple wall connections to framing above*

Detail F1—Joists Perpendicular to Wall

If a rim joist was used instead of the blocks shown, L90 or equivalent connectors may be used at greater spacing than the L50s shown.

For stitch-nailing between top plates, see Details W1 and P1

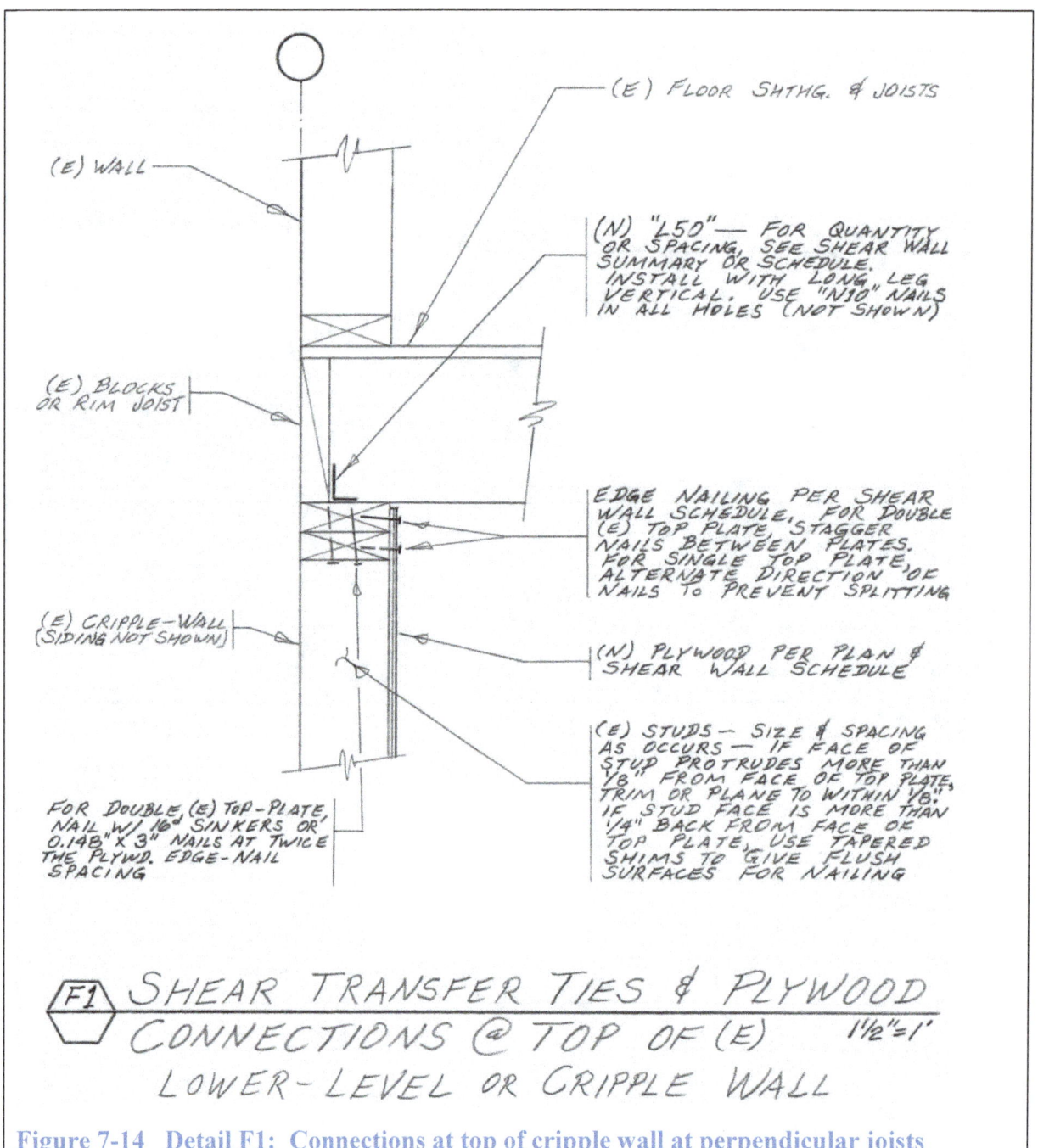

Figure 7-14   Detail F1: Connections at top of cripple wall at perpendicular joists

Detail F2—Joists Parallel to Wall

For stitch-nailing between top plates, see Details W1 and P1

Blocking at last joist bay keeps the end joist from rolling over, which can occur when several thousand pounds of force are applied along the joist and the whole house is shaking severely.

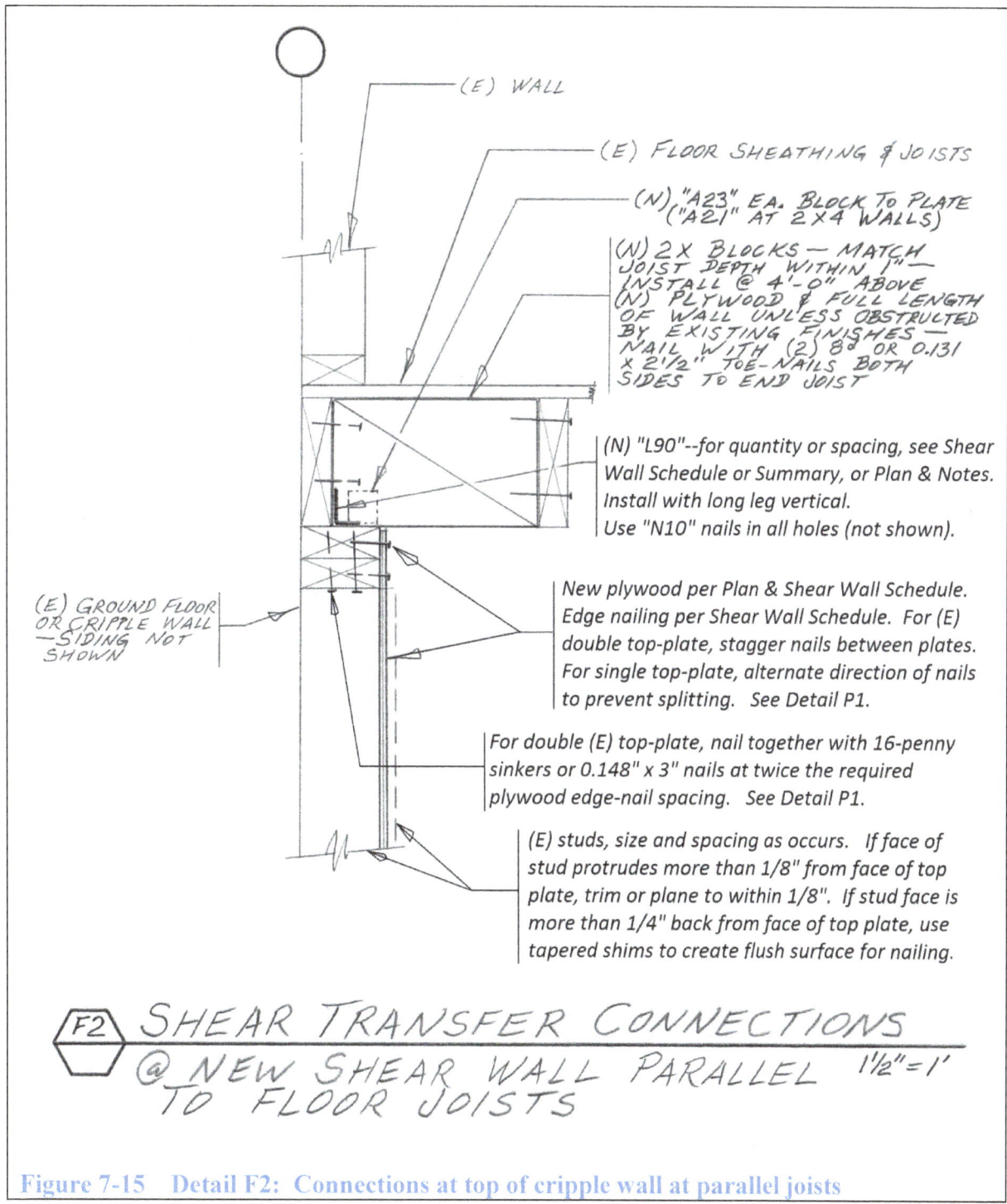

Figure 7-15   Detail F2: Connections at top of cripple wall at parallel joists

Detail F3—Cripple Wall Without Top Plate

Use adhesive between existing end joist and new filler members; *Do not* use adhesive to attach shear panels.

Blocking at last joist bay keeps the end joist from rolling over, which can occur when several thousand pounds of force are applied along the joist and the whole house is shaking severely.

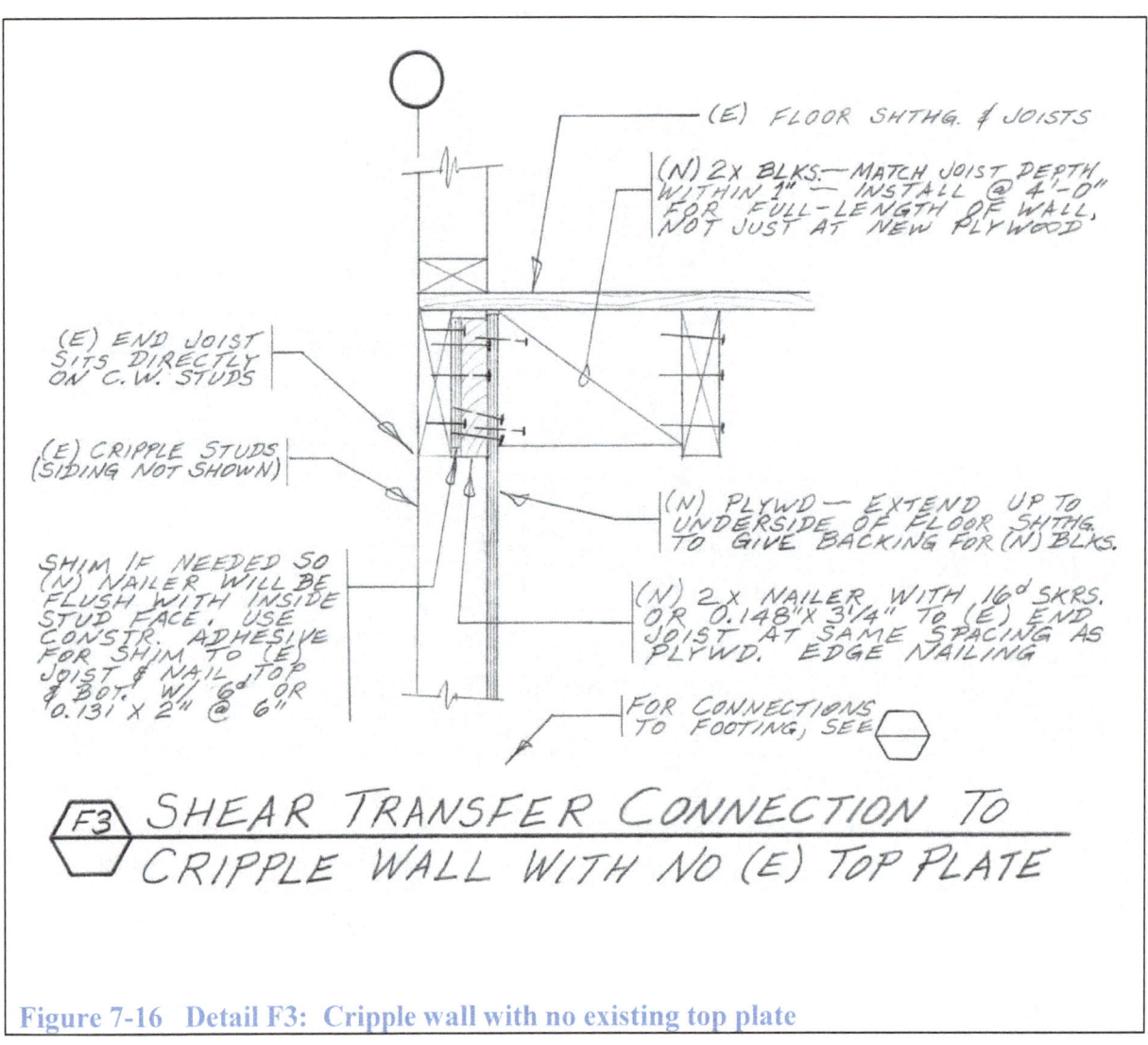

Figure 7-16   Detail F3: Cripple wall with no existing top plate

Detail F4—Connections to Modified Balloon Framing

For the Stud Blocking Alternative, 3x4 blocks are preferred over 2x4 blocks to reduce the tendency of the blocks to split.

Structural screws from top of joist to cripple wall top plate are the same for either alternative.

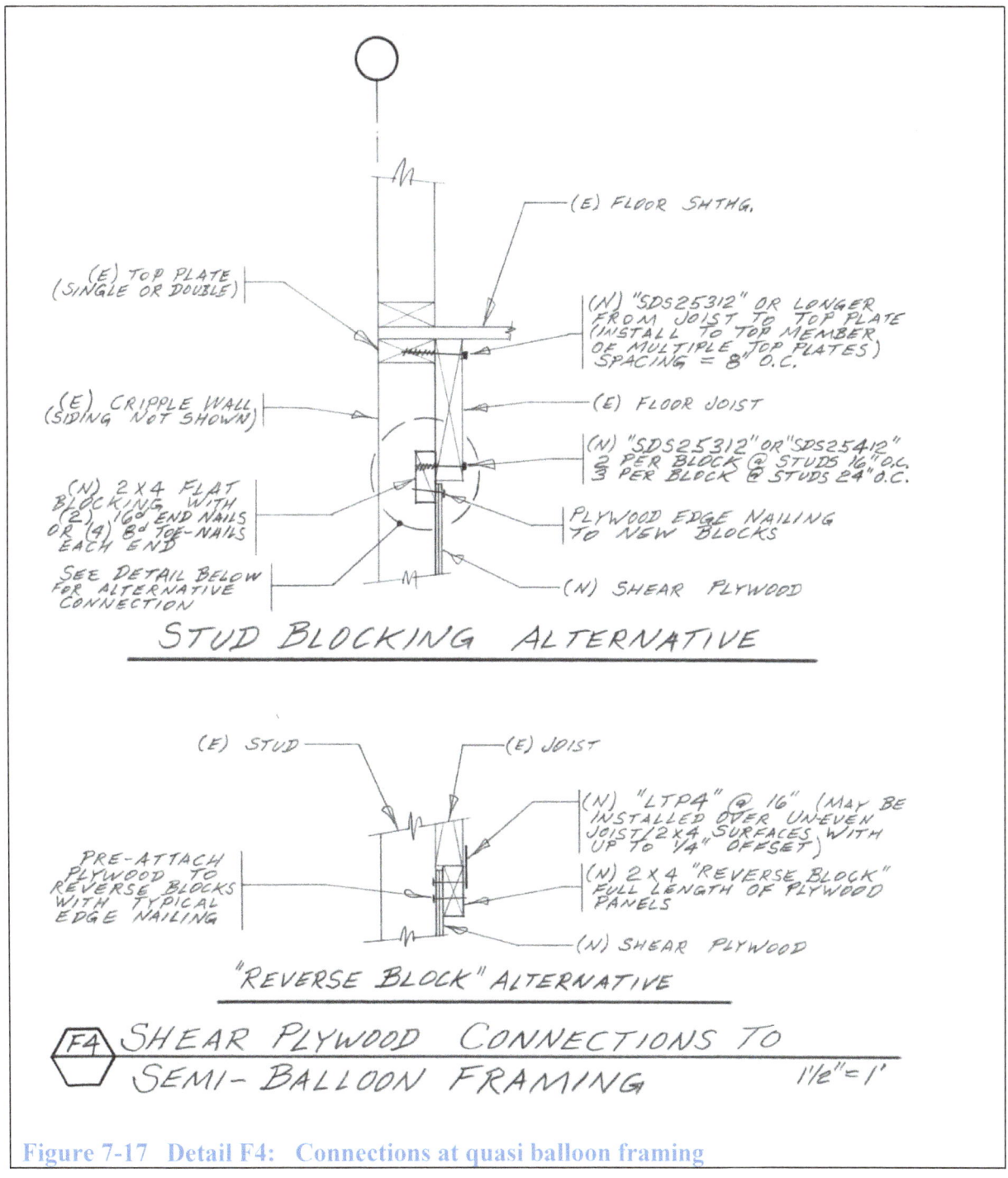

Figure 7-17 Detail F4:   Connections at quasi balloon framing

Detail F5—Balloon Framing with Joists Perpendicular to Wall

If you can establish that the existing fire-blocking is installed with two nails into the top of the joist, the new blocks may be omitted and the plywood nailed to the fire-blocks.

This detail relies on the floor sheathing to provide "diaphragm chord ties." If the ledger beneath the floor joists is continuous (or joints are spliced with appropriate means) it may also function as a chord tie. Best practice would be to install blocks between the floor sheathing and the top of the ledger. [Remember that blocks cost about $50 apiece to install.]

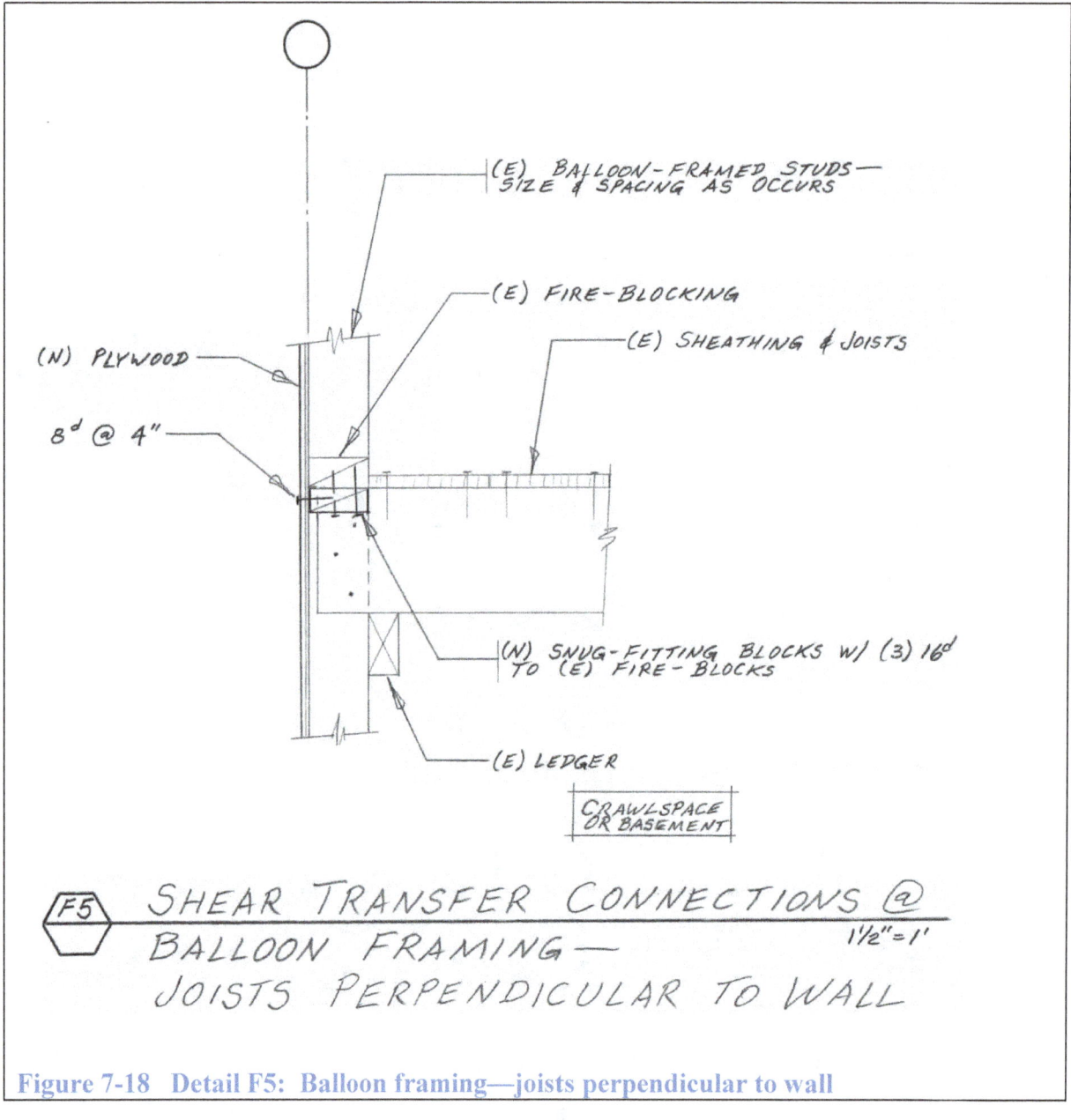

Figure 7-18 Detail F5: Balloon framing—joists perpendicular to wall

Detail F6—Balloon Framing with Joists Parallel to Wall

If the existing fire-blocking is snug-fitting and the joist is nailed to each stud with at least two nails, nailing the plywood to the fire-blocking may provide an adequate connection.

Any breaks in the joist should be spliced in order to provide diaphragm chord ties.

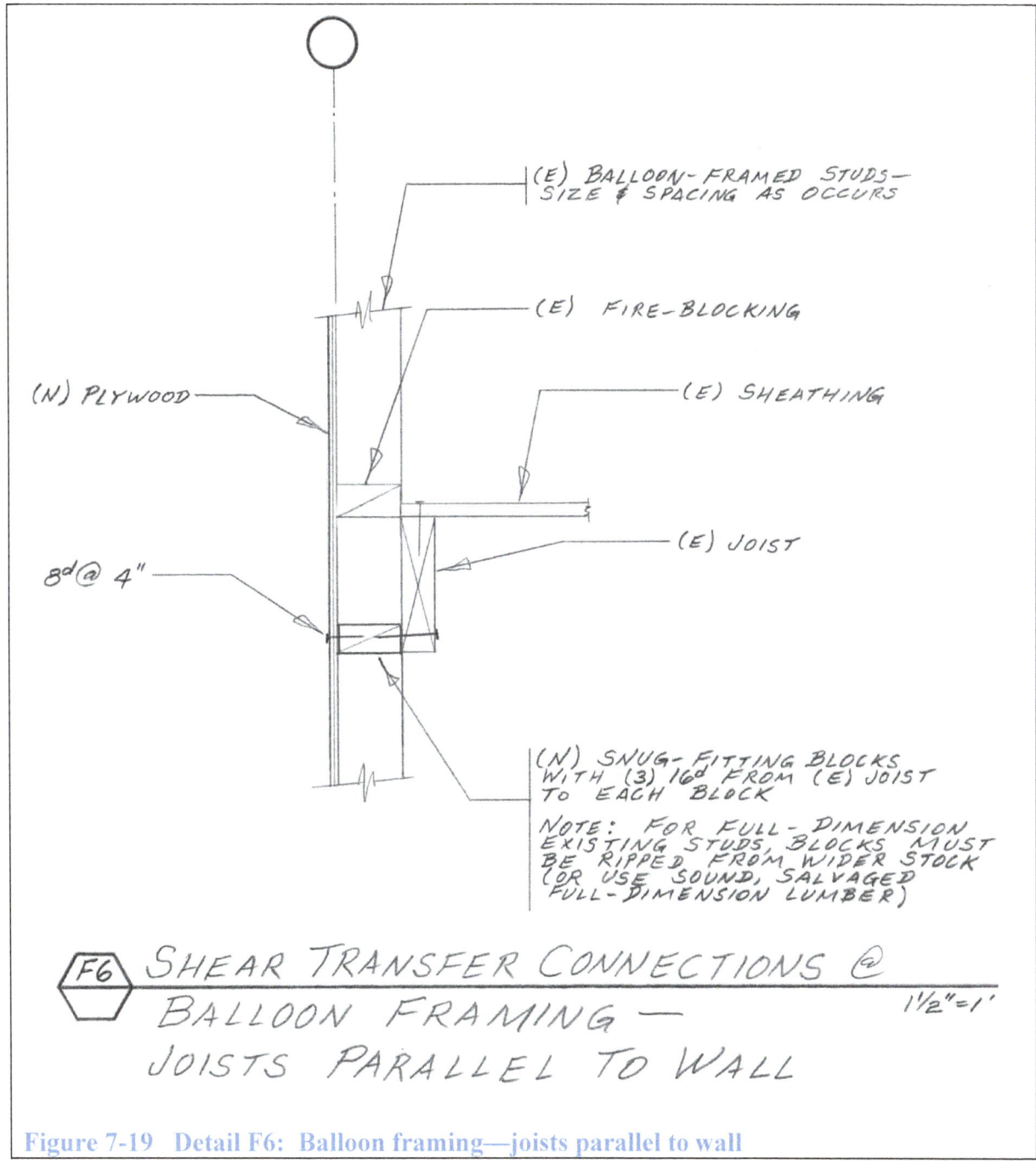

Figure 7-19   Detail F6: Balloon framing—joists parallel to wall

Detail F7—Split-Level Floor Framing—Joists Perpendicular to Walls

NOTE: Unless the floor diaphragms at both floor levels are stabilized on all four sides by shear walls or other means, further connections must be installed to keep the building segments from moving apart. Also this detail relies on floor sheathing boards for "diaphragm chord ties" at the upper floor level. Best practice would be to install a continuous tension member (such as a steel strap) along the bottom of the joist blocks at the upper floor.

This detail shows platform framing at the lower level and balloon framing at the upper level. Conditions may vary. Likewise, joist direction may change from one level to the next. You may need to mix-and-match connections shown here and in Detail F8.

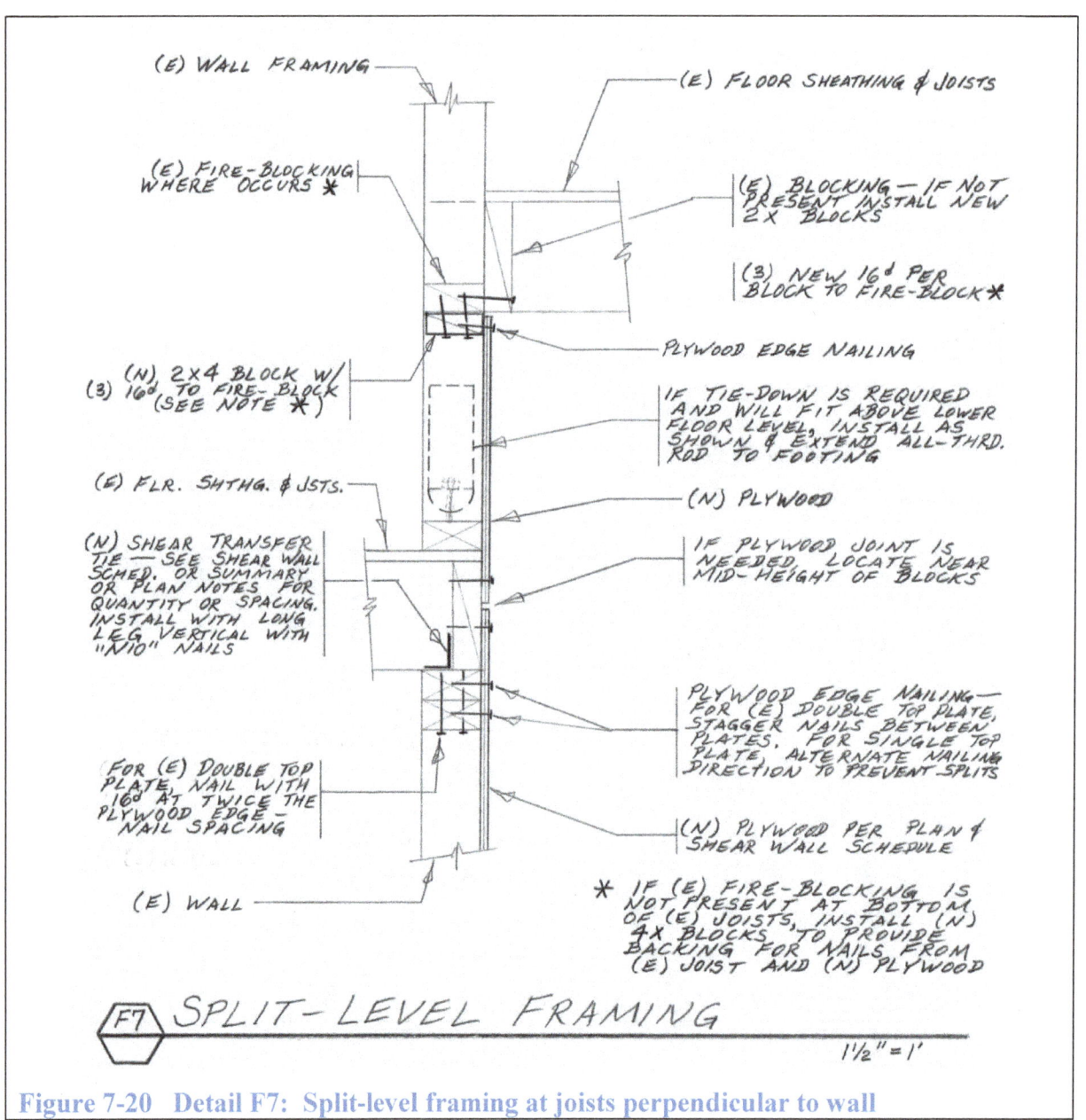

Figure 7-20   Detail F7: Split-level framing at joists perpendicular to wall

Detail F8—Split-Level Framing—Joists Parallel to Wall

NOTE: Unless the floor diaphragms at both floor levels are stabilized on all four sides by shear walls or other means, further connections must be installed to keep the building segments from moving apart.

Joist direction may change from one level to the next. You may need to mix-and-match connections shown here and in Detail F7.

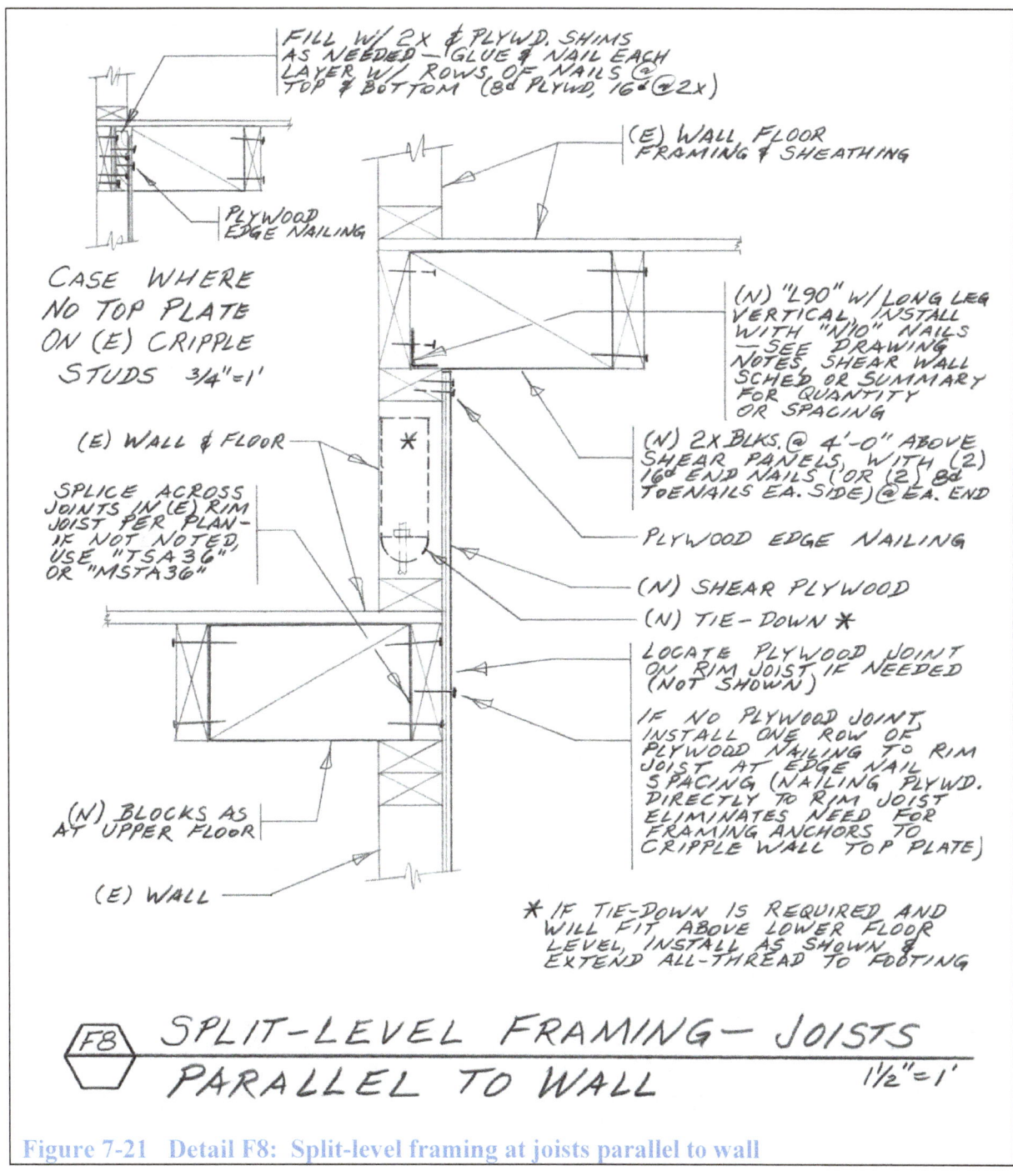

Figure 7-21 Detail F8: Split-level framing at joists parallel to wall

Detail F9—Retrofit Joist Blocking

Several methods to provide new blocking for shear transfer from floor diaphragm to shear panels below.

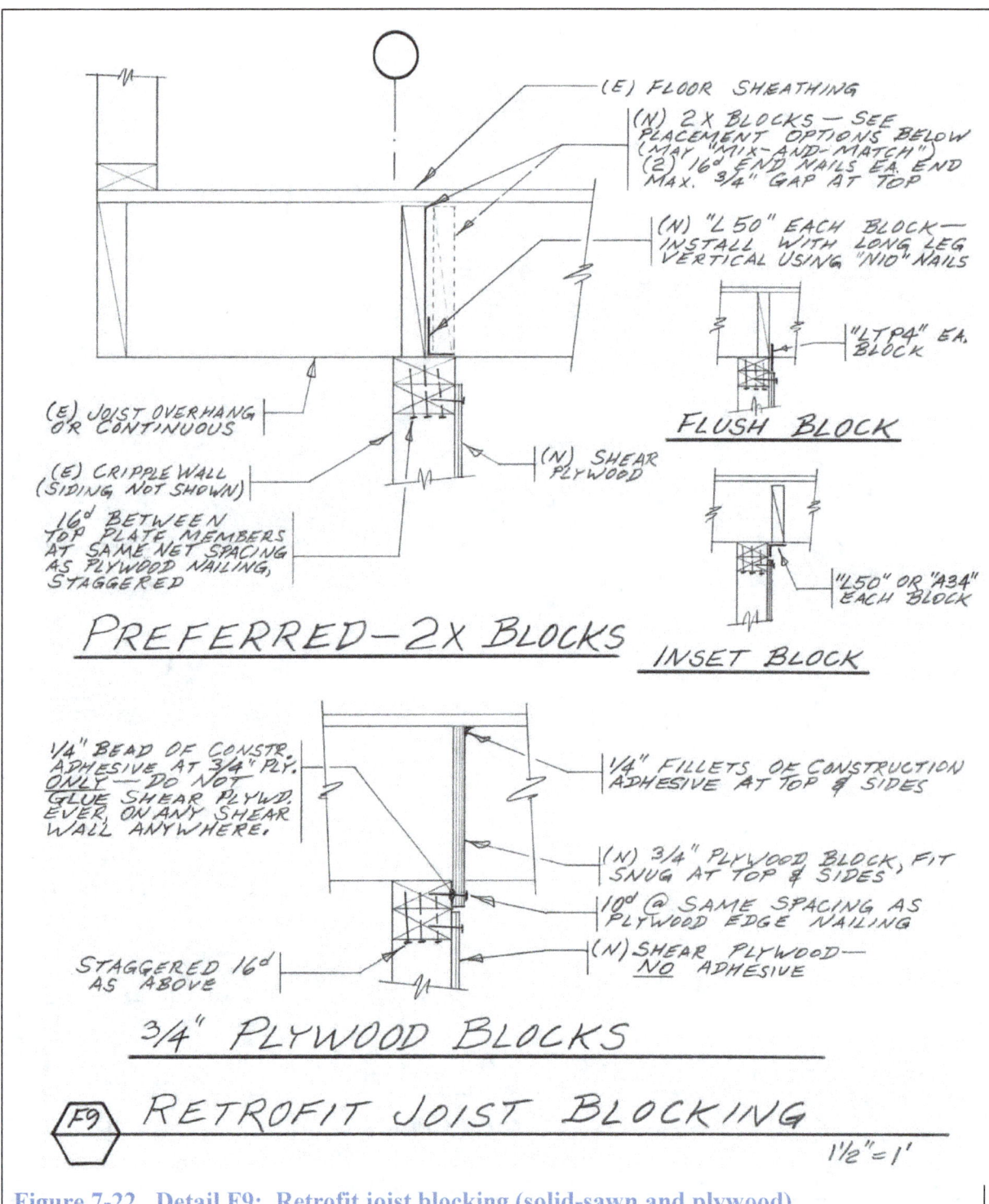

Figure 7-22   Detail F9: Retrofit joist blocking (solid-sawn and plywood)

7.3 Shear transfer diaphragms

Detail D1—Shear Transfer Diaphragm Connection to Foundation Wall

This detail is similar to the "reverse block" connection method, but with the plywood oriented horizontally.

MSPs used to attach the nailer to the foundation reduce the number of bolts needed.

If the original floor sheathing is plywood use a longer shear transfer diaphragm, or attach it to two floor joists, to make up for the likely case that the plywood is nailed to the joists at 10" (rather than 6" along the edge connected to the mudsill).

Plain lumber may be used for the nailer if "peel-and-stick" membrane is used between it and the concrete.

For view looking up at plywood, see Detail D2.

Figure 7-23 Example of a shear transfer diaphragm connecting first joist to full-height foundation stemwall. This photo shows the connection illustrated in Figure 7-24. Black material is self-adhered flashing ("peel-and-stick" flashing) used to isolate the untreated wood from the concrete instead of using pressure-treated material.
For other photos of shear transfer diaphragms see Figure 6-62

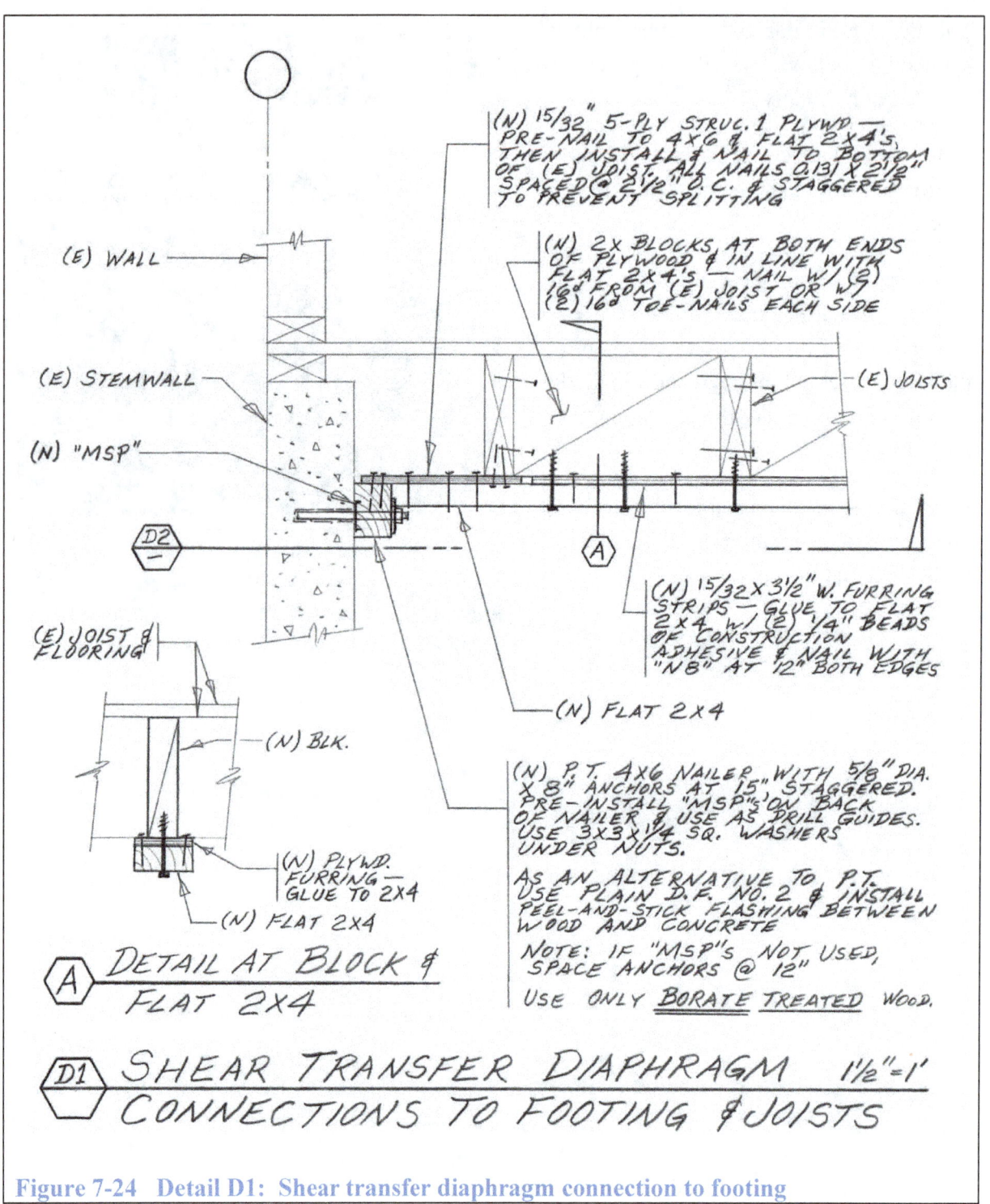

Figure 7-24 Detail D1: Shear transfer diaphragm connection to footing

Detail D2—Shear Transfer Diaphragm View Looking Up

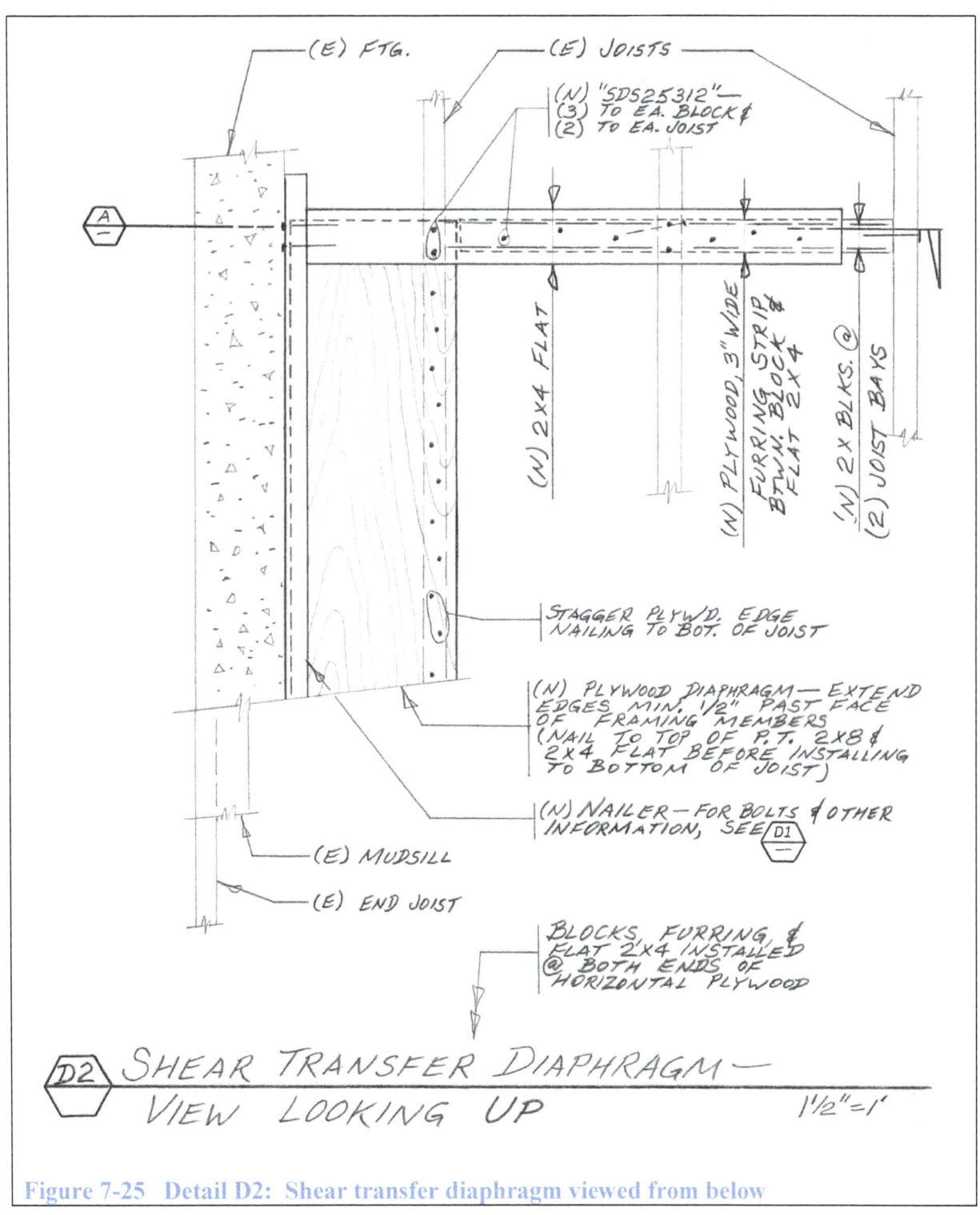

Figure 7-25 Detail D2: Shear transfer diaphragm viewed from below

Detail D3—Shear Transfer Diaphragm Connection to Wood-Framed Wall

This detail is used when STTs cannot be used at the end joist, or there is not enough access to install them.

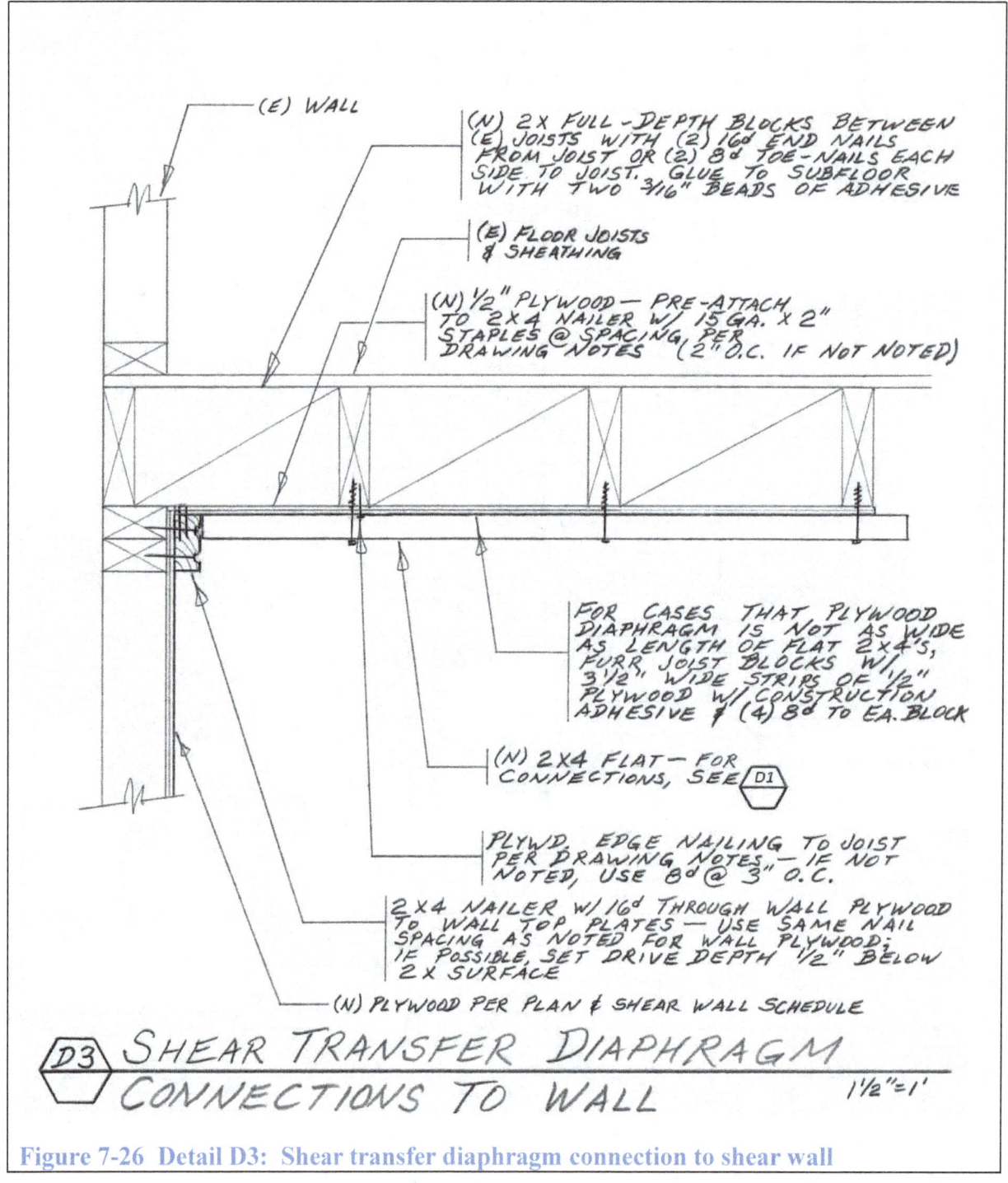

Figure 7-26 Detail D3: Shear transfer diaphragm connection to shear wall

7.4 Collectors

After wall top-plates, joist connections are commonly supplemented to create collectors & drag ties. Collectors must resist tension or compression. Collector design can quickly get complex.

Details T1 & T2—Reinforcement of Existing Joist Lap

Floor joists often lap together on top of a center beam or wall. To provide compression capacity an additional "bearing block" from a length of 2x stock is butted against the end of one of the joists (which ever end is most accessible). The new 2x member is securely connected to the adjacent joist.

To provide the needed tension capacity a new steel strap is connected across the joint from the original joist to the new bearing block.

Note that the detail requires more nails to connect the 2x8 to the original joist than there are connecting the strap. This is because nails have lower capacity when connecting wood-to-wood than steel-to-wood.

Figure 7-27 Real life installation shown in Details T1 and T2, shown in Figure 7-28. (Note: Installation is a mirror-image of that shown in details, and uses structural screws to attach the 2x8 instead of nails.)

Details T1 and T2—Reinforcement at Joist Lap for Added Drag Capacity

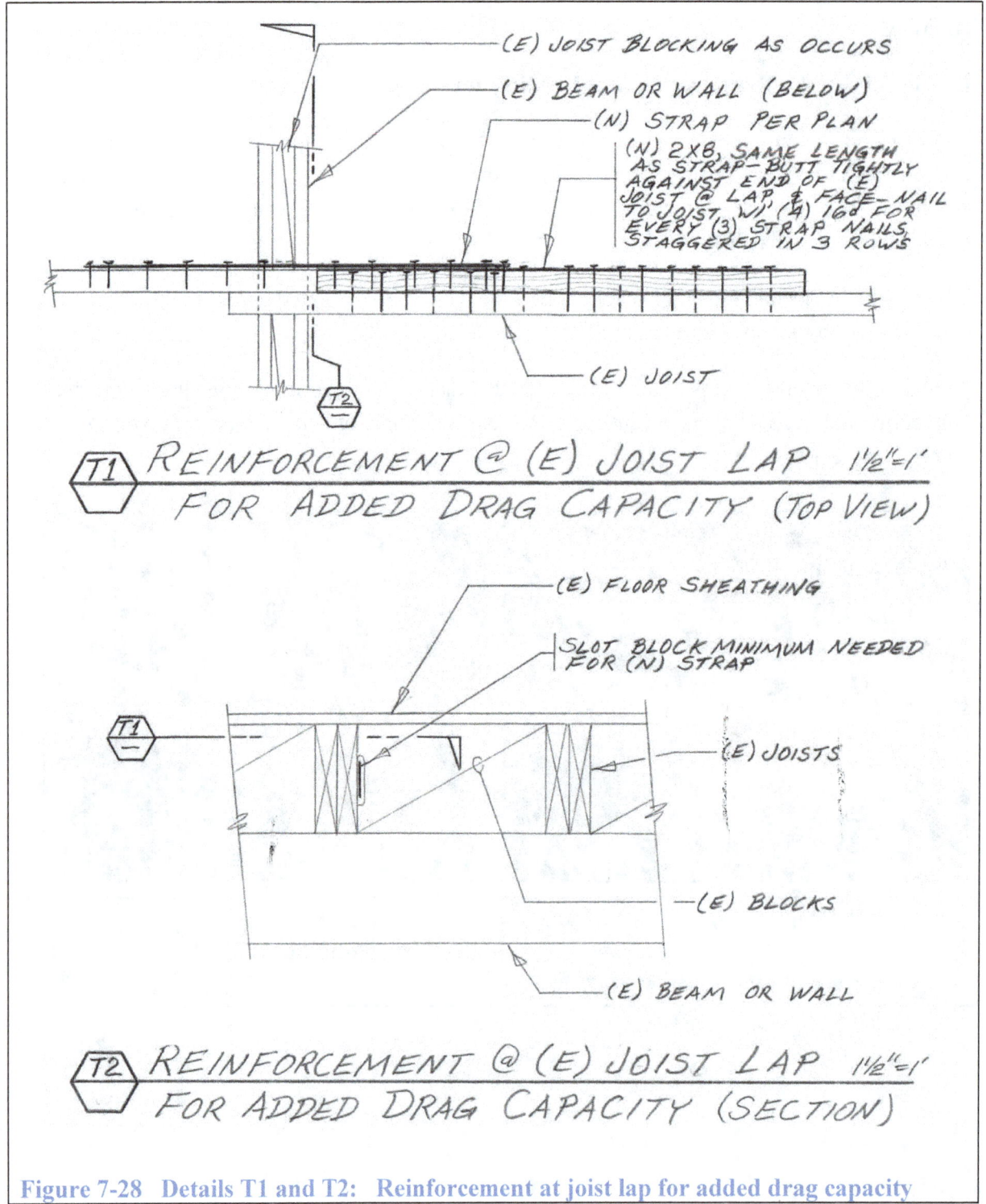

Figure 7-28   Details T1 and T2:   Reinforcement at joist lap for added drag capacity

Detail T3—Drag-Tie/Collector

Used where a short length of foundation is available in line with the collector, but not enough for sufficient anchorage using UFP10s or similar.

New 4x8 drag-tie connects to new LVL cleats that are in turn connected to the underside of existing joists. 4x8 bears against footing beyond to resist compression; all-thread rod from horizontally-oriented tie-down extends back to another tie-down that is bolted to the footing.

For side view, see Detail T5

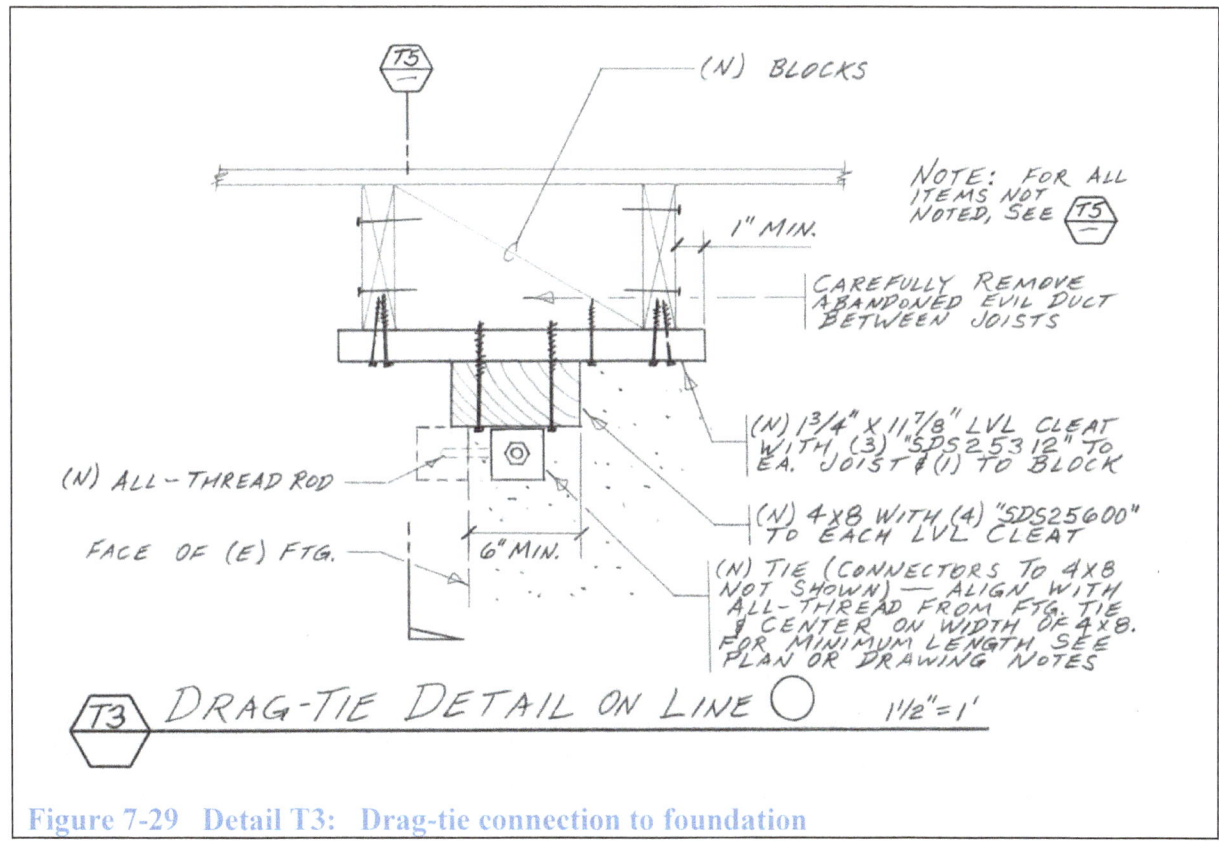

Figure 7-29 Detail T3: Drag-tie connection to foundation

Detail T5—Drag-tie/Collector

Side view of collector shown in Detail T3. Note that LVL cleats can be located to leave room for existing pipes, conduits, or electrical cables.

Bracket to footing is typically an HDB-type "hold-down" with expansion bolts to concrete; bracket to 4x8 is an HDU or similar.

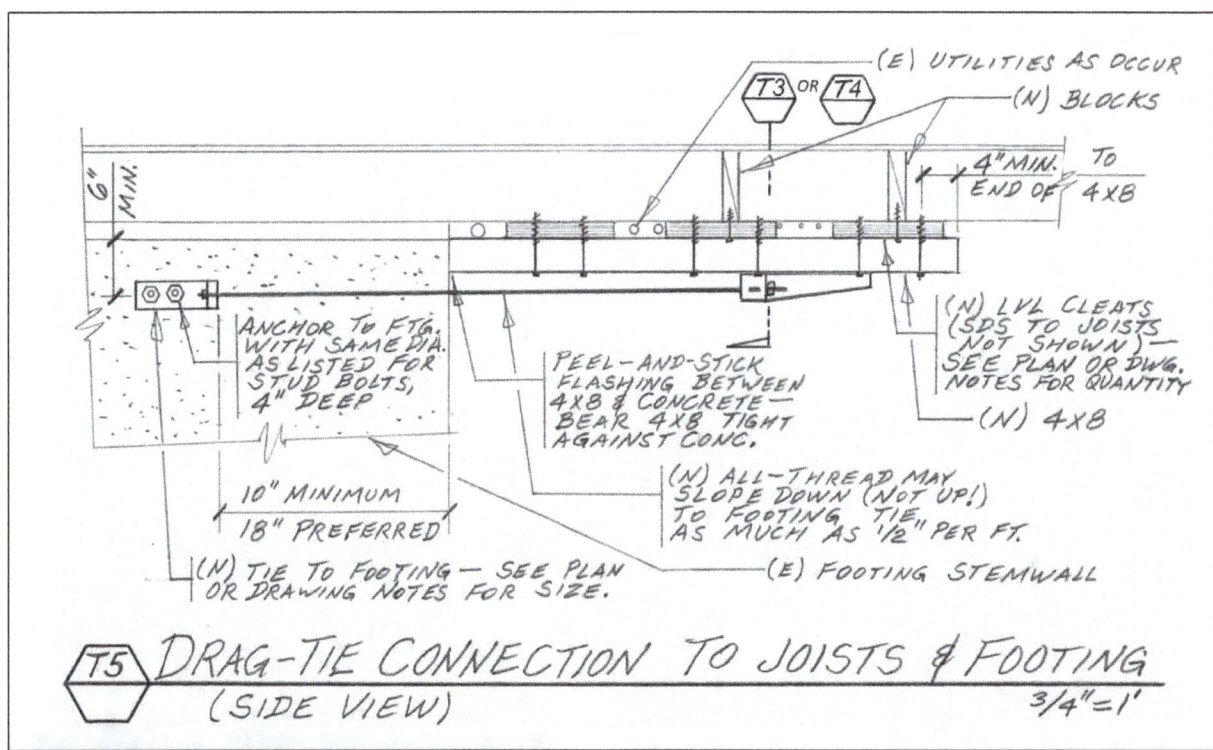

Simpson "HD7B" attached to face of concrete foundation stemwall. All-thread rod extends to bracket connected to underside of floor joists.	LVL cleats attached to underside of original floor joists. 4x6 connected to cleats distributes loads between the cleats. "HDQ8" bracket connects to all-thread rod extending back to bracket shown in photo at left.
Photos: Julio Nieto, Nieto-Valle Construction (formerly Anderson-Niswander)	
Figure 7-30 Detail T5: Drag-tie connection to joists & footing (side view)	

Detail T6—Drag-tie/Strut at Fireplace

Fireplaces are often set toward the outside of a wall, so the inside face of the brick is flush with the inside wall surface. Wood-framed walls begin on either side of the brick fireplace/chimney column. Connecting the wood-framed walls on both sides of the fireplace helps hold the building together in an earthquake (until the chimney collapses—but I told you to take it down anyway).

Often the brickwork in the crawlspace was not as uniform and smooth as the work exposed in the living room above, so the 4x4 strut shown has to be spaced out away from the cripple wall to provide clearance.

For photos of this connection see Figure 6-39

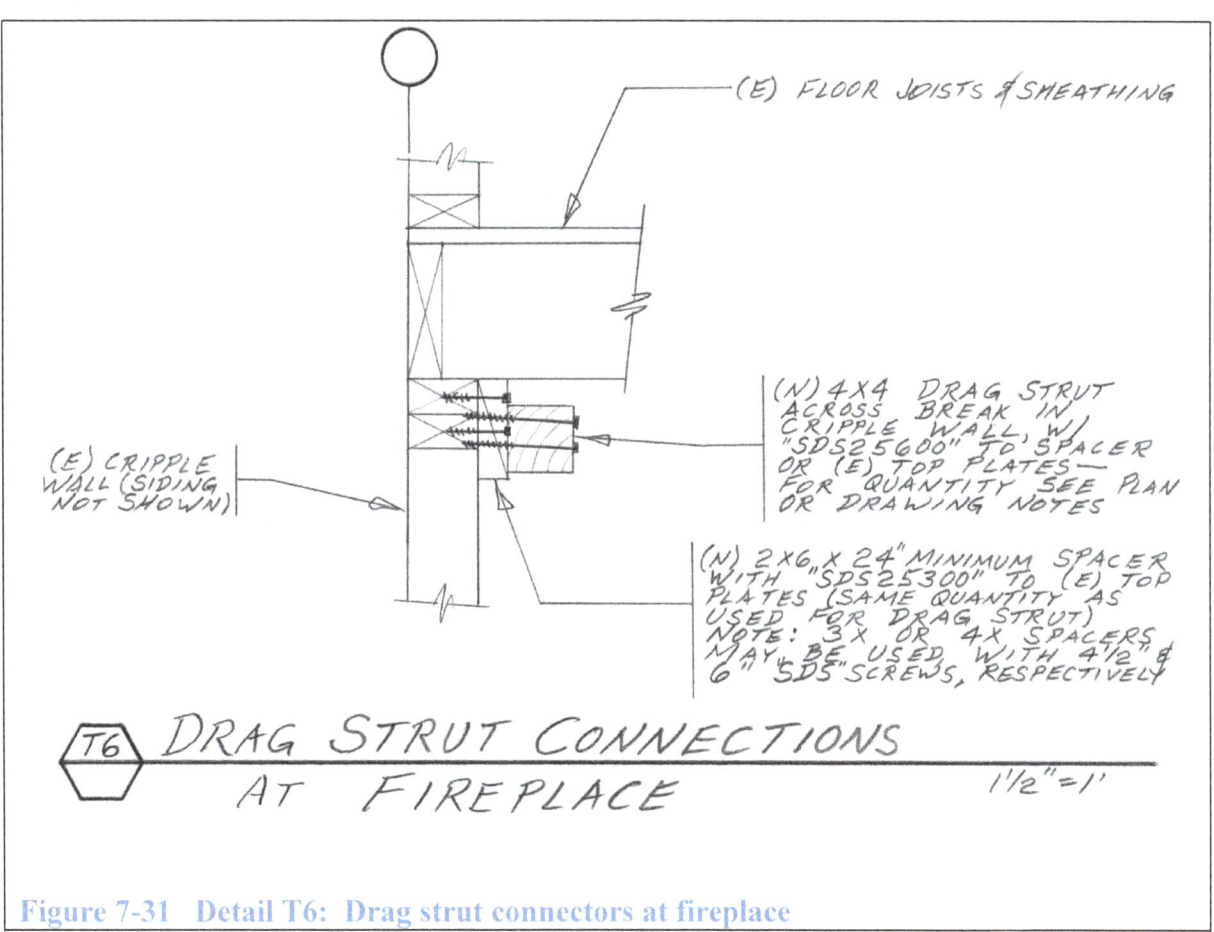

Figure 7-31 Detail T6: Drag strut connectors at fireplace

Detail T7—Drag-Strut at Inside Corner

Sometimes the wall jogs out (toward the bottom of the page in the drawing shown) to form a bay or other floor extension, and you have to connect floor joists at the extension to the wall (on the right in the drawing).

This connection is essentially the same as Detail T1.

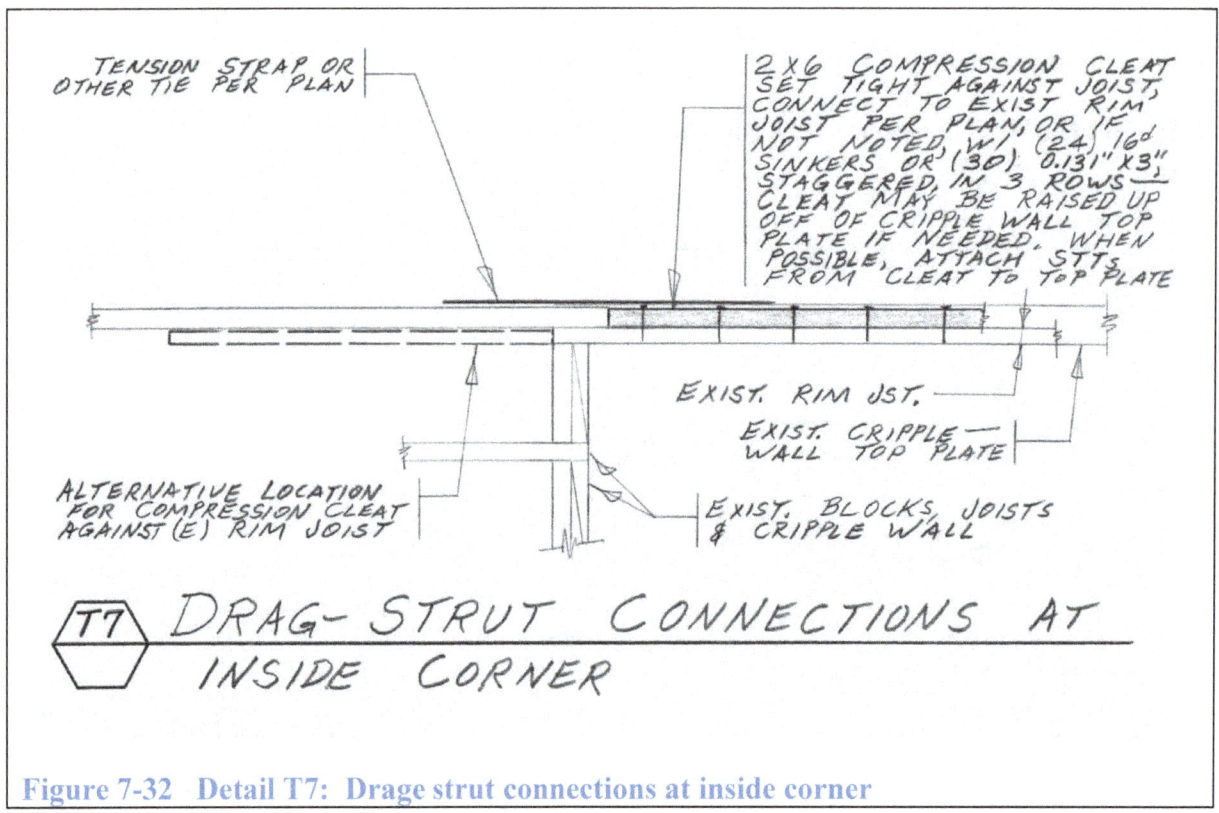

Figure 7-32   Detail T7: Drage strut connections at inside corner

Detail T8—Tie-Rod Through Framing with Blocking

This detail allows new ties to be concealed above a finished ceiling. The all-thread rod can be angled slightly to avoid existing utilities within the floor framing system.

Best practice is to install the all-thread and blocks as close together as possible, and as straight as possible, with blocks fit tightly in place and no slack in the rod connections.

See Figure 7-36 for photos of installation similar to that shown in Detail T8

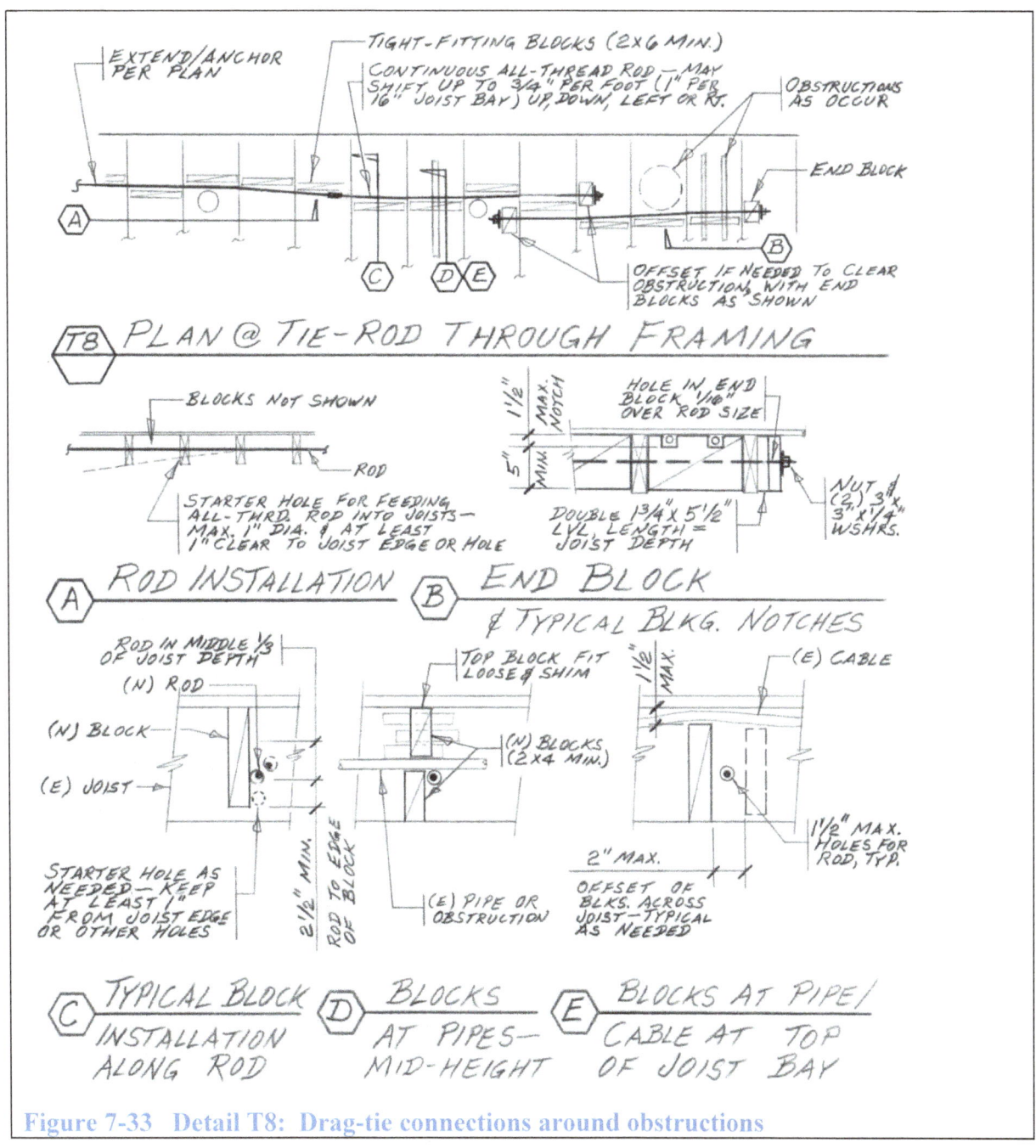

Figure 7-33   Detail T8: Drag-tie connections around obstructions

Details T10 and T11: Drag Strut Connections

An installation somewhat similar to Detail T11 is shown in Figure 6-39

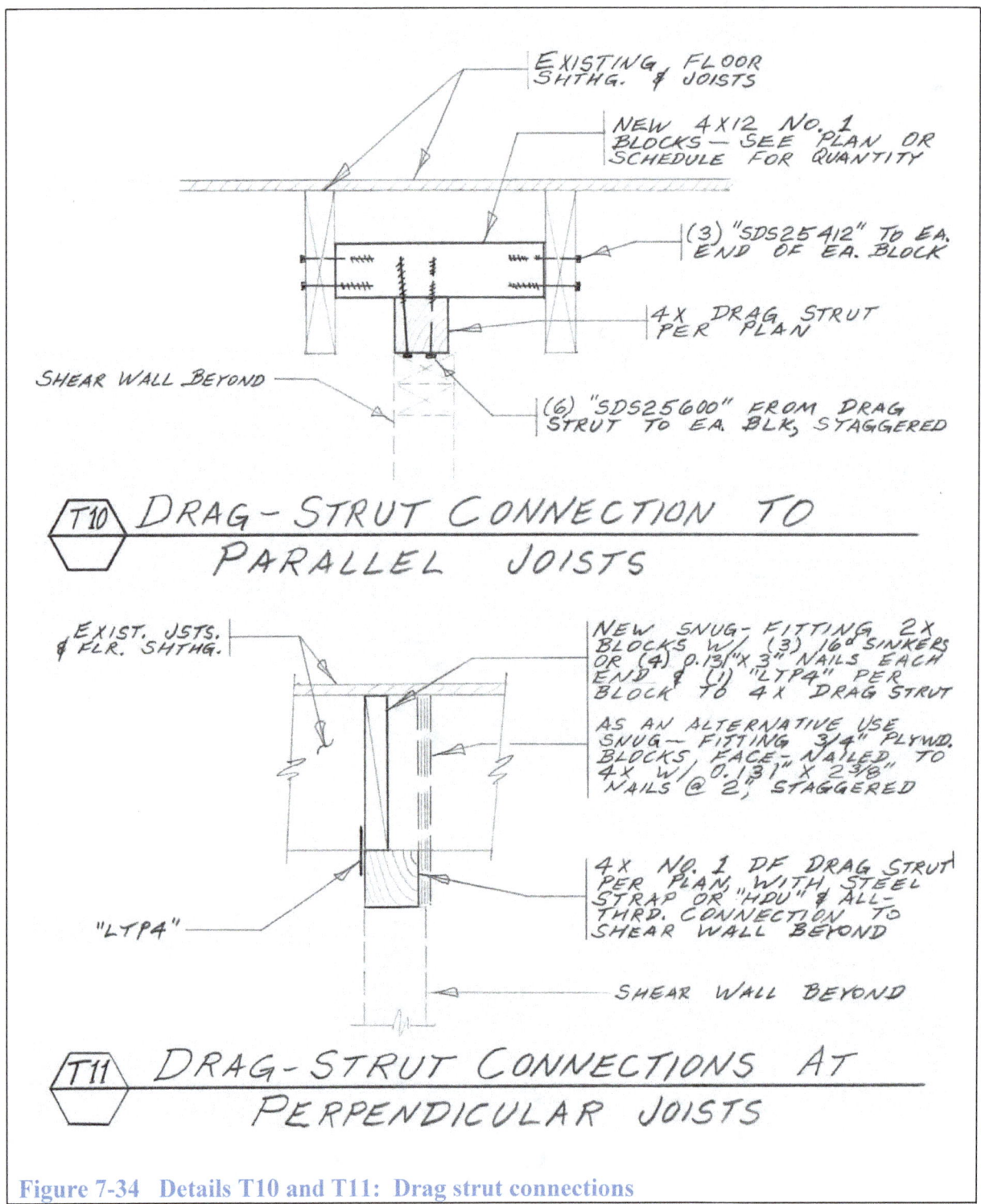

Figure 7-34   Details T10 and T11: Drag strut connections

Collector Connections

The "*LVL extended from foundation beyond*" is the same LVL shown side-bolted to the foundation in Detail B3 (Figure 7-10). In this case the foundation stepped down from the full-height condition in Detail B3 to the case shown here with a cripple wall. See Figure 7-37 for photos of this installation.

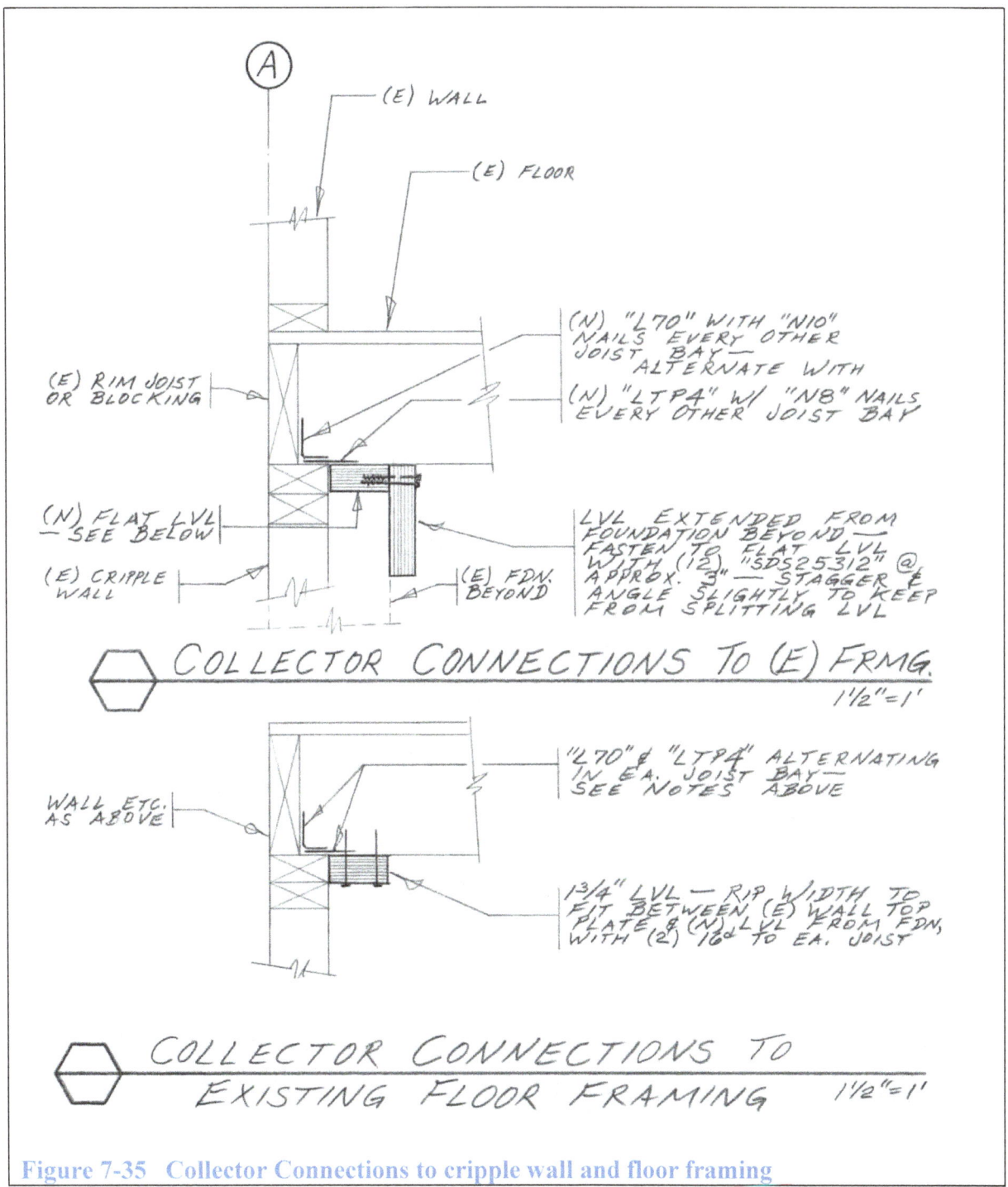

Figure 7-35 Collector Connections to cripple wall and floor framing

Bearing block at end of all-thread tie (viewed looking up).
New joist block resists compression; all-thread joined with coupling nuts (visible at bottom of photo) resists tension.

LVL at left side of photo connected to original cripple wall top plate with LTP4s

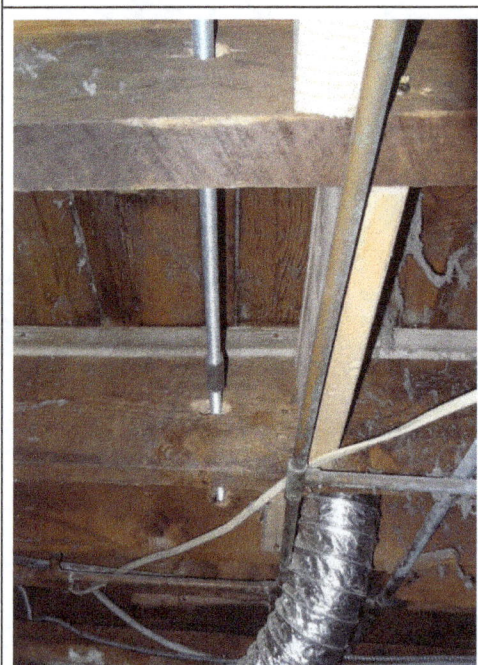

Joist bocks and tie-rod extended where multiple obstructions would prevent installing a strut on the underside of the joists.

Transition from LVL under joists to wider LVL "side-bolted" to foundation stemwall. Note structural screws between LVLs

LVL connection to foundation stemwall.

Figure 7-36 Photos above show all-thread tie similar to that shown in details from Figure 7-33.

Figure 7-37 Photos above show real life installation for details in Figure 7-35.

7.5 *Plywood/miscellaneous*

T1-11 Plywood attachment detail

Grooved plywood siding was rarely installed correctly to resist earthquake forces. Your choices for adding more nails will vary depending on where the studs occur behind the plywood panel joints.

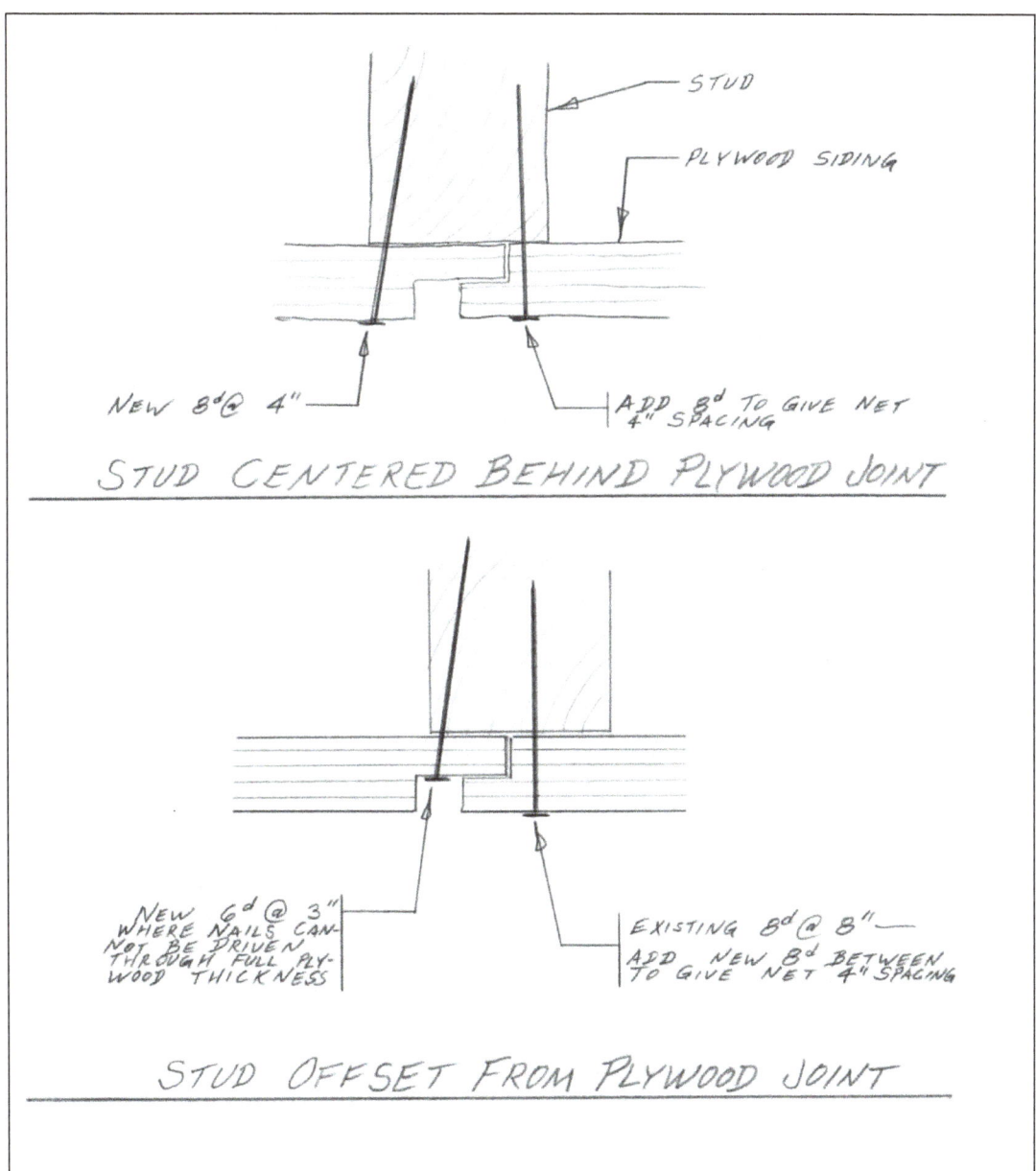

Figure 7-38—Grooved plywood nailing upgrades. Some patterns of plywood grooves will not allow nailing through the full panel thickness—in such cases the bracing strength is based on the panel thickness through which the nails are driven.

Detail P2—Plywood Connections at Wall Intersections

It is often easiest to remove non-bearing walls that run into a wall that must receive shear panels. In cases where this is not practical, some other options are shown.

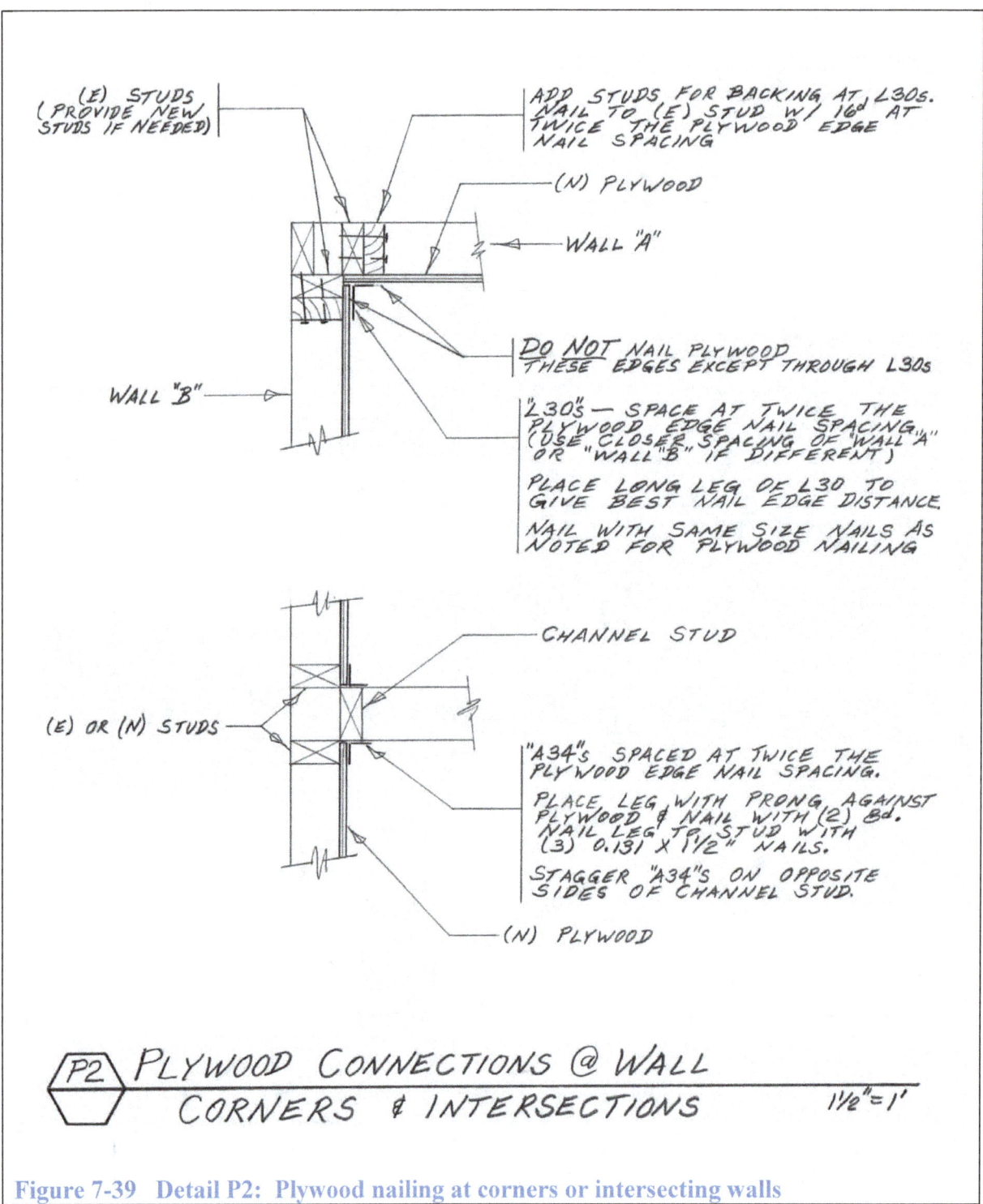

Figure 7-39   Detail P2: Plywood nailing at corners or intersecting walls

Detail—Plywood splice at intersecting wall

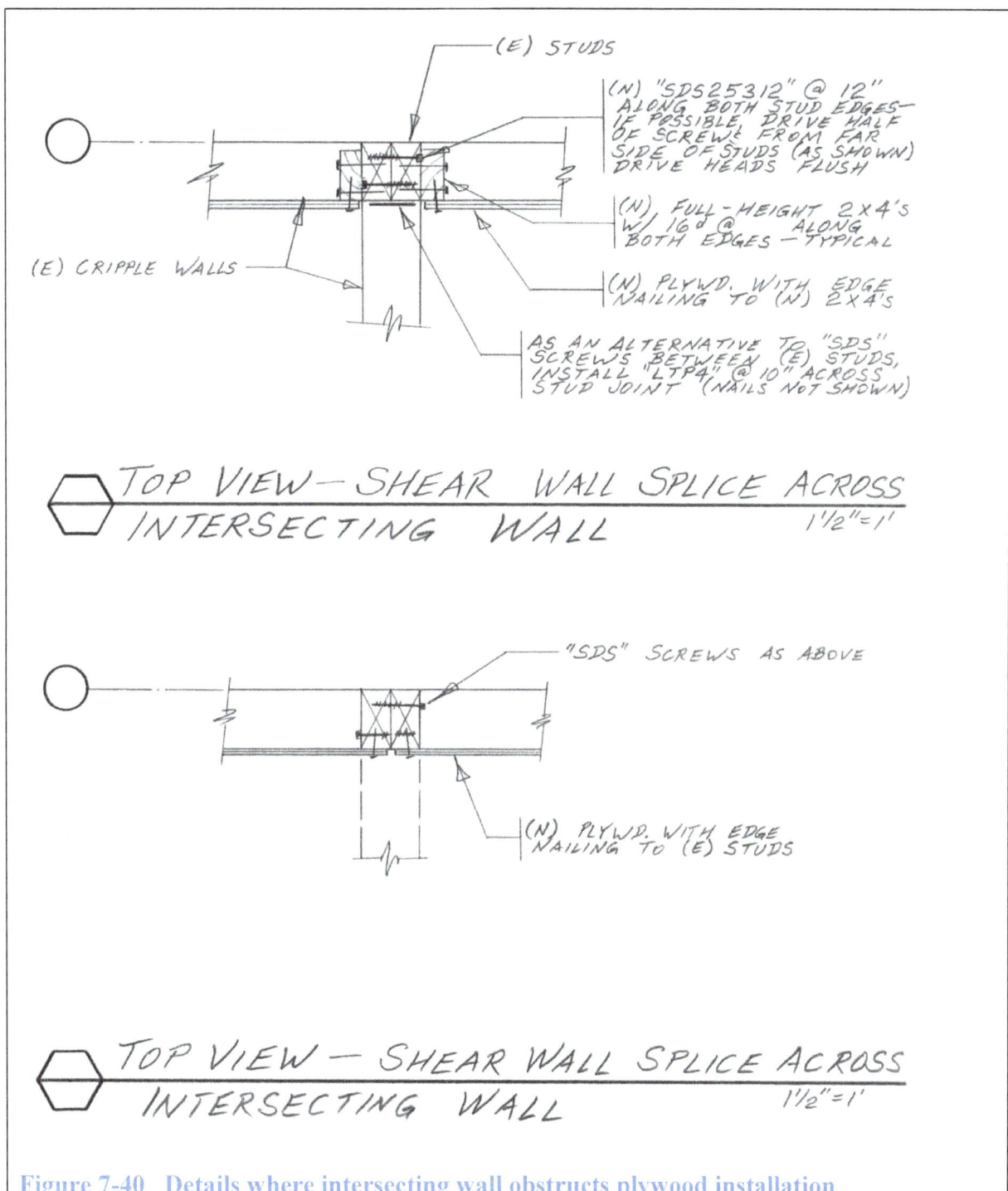

Figure 7-40   Details where intersecting wall obstructs plywood installation

Detail P1—Plywood Nailing Explanations

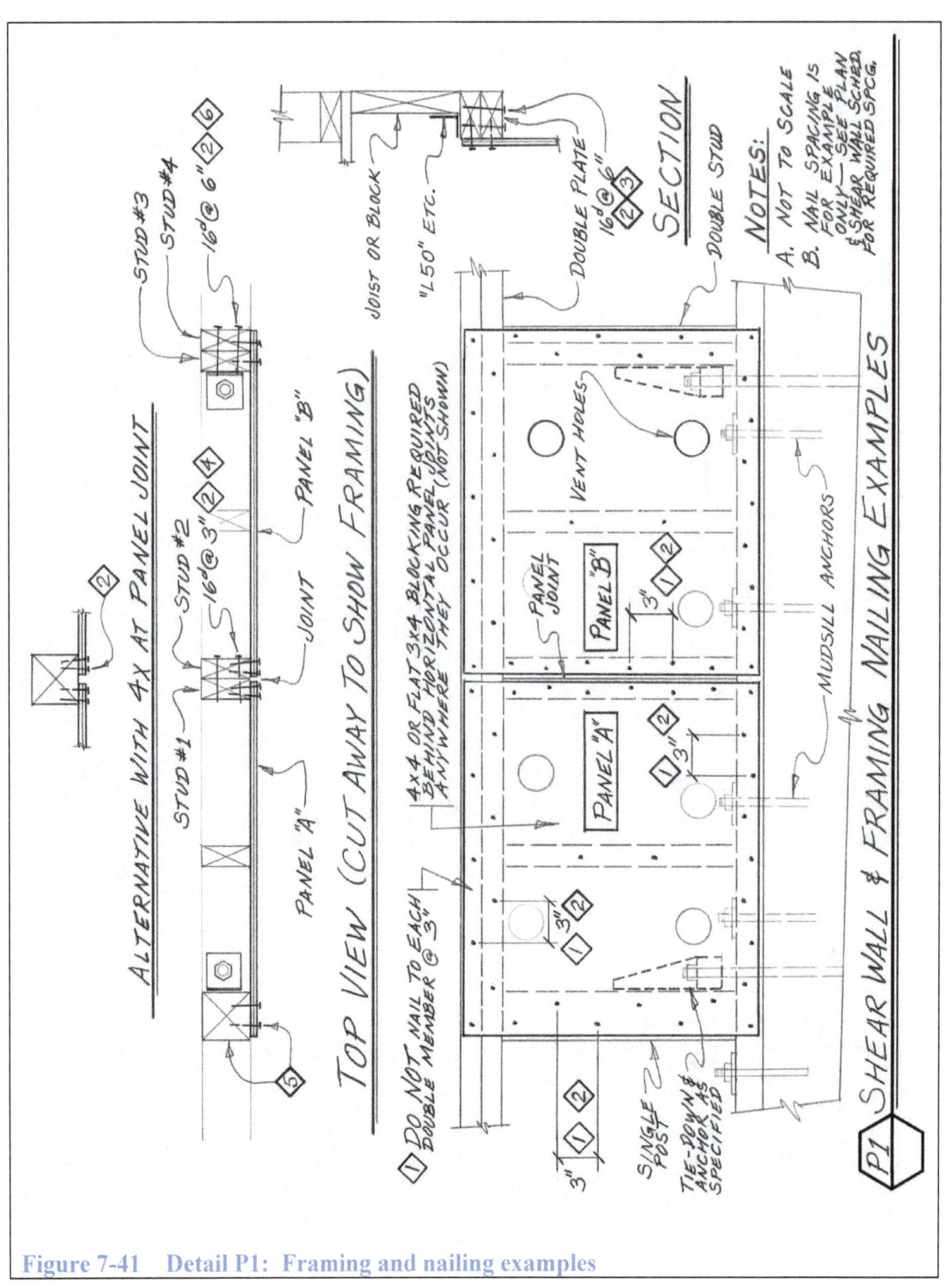

Figure 7-41   Detail P1: Framing and nailing examples

Notes for Plywood Nailing Explanations

1. Panel edge-nail spacing should be the same along all edges of every piece of plywood. 3" edge nailing is shown as an example in the illustration above. The plywood knows only what the edge-nail spacing is—it does not know whether the nails attach to a single 4x4 post, to sistered studs, or a double top-plate. If there are TOO MANY NAILS along one edge of the plywood, then the nailing along the other edges of the panel will fail first. This is not good, so it is important to have equal nail spacing along all panel edges.

2. "Staggered nailing" is intended to reduce the chance of splitting framing members that could happen if the nails are all driven in a straight line. "16-penny at 3 inches, staggered" means the nail spacing is 3 inches, but the nails alternate back and forth. This is the same as installing two rows of nails at 6-inch spacing, etc. In the case of edge nailing to a double end stud or top plate, alternate the nails from one member to another. Nailing to double studs at panel joints is at 3" along both panel edges.

3. Earthquake (EQ) force is transferred into double top plate through the "L50" framing anchor or other connection from the blocking or rim joist. Half of the EQ force is immediately transferred from the upper top plate member into the shear plywood through the panel edge nailing. The other half of the EQ force must transfer into the lower member of the double top plate, and then into the shear plywood through the edge nails driven into the lower member. Therefore the "stitch nailing" between the top plate members can be at twice the spacing of the panel edge nailing (in the figure, 6" O.C.)

4. For doubled studs behind panel joint: ALL of the force in Panel "A" must transfer across the joint between the studs and into Panel "B". Therefore the stitch-nailing between studs must be at the same spacing as the shear panel edge-nailing. In the example, Stud #1 must be nailed to Stud #2

5. For single end-post with hold-down, all the plywood edge nails connect to the same wood member that the hold-down attaches to. No special requirements.

6. For built-up studs connected to hold-down: Multiple studs must be connected to act as a single member. In the example, half of the plywood edge nails connect directly to the stud attached to the hold-down (Stud #3). This transfers half of the EQ force from the plywood to the hold-down. The rest of the EQ force goes from the plywood into Stud #4, so Stud #4 must connect to Stud #3 with one 16-penny nail for every plywood edge nail (in this case, 6-inch O.C. nailing between Studs 3 & 4).

NOTE—Nail spacing between Studs 3 & 4 is twice the spacing required between Studs 1 & 2. If there is any doubt about the reason for this, the best thing is to nail ALL built-up studs together with 16-penny nails at the same spacing as the edge nailing—that way you can be sure you always have enough nails connecting the studs. Nailing between studs is different than plywood edge-nailing: unless you start to split the studs, you can't have "too many" nails. You can have "too many" plywood edge nails, as noted in Item 1 above.

7. Stitch-nailing between double plates and studs is based on 16-penny COMMON nails, and is not precisely engineered. For Sinkers or 0.148" gun nails, add one nail for every three.

8. Splice ALL joints in double top plates of cripple-walls with (24) 16-penny nails on each side of joint, whether or not the splices occur in areas that will receive plywood sheathing with 16-penny nails at 3" O.C.

9. Vent holes are not required when panels will be installed on the conditioned side of walls in living areas, or if the panels will be covered with other wall finish materials, or where dampness will not be a concern.

Detail X1—Lock Block Installation

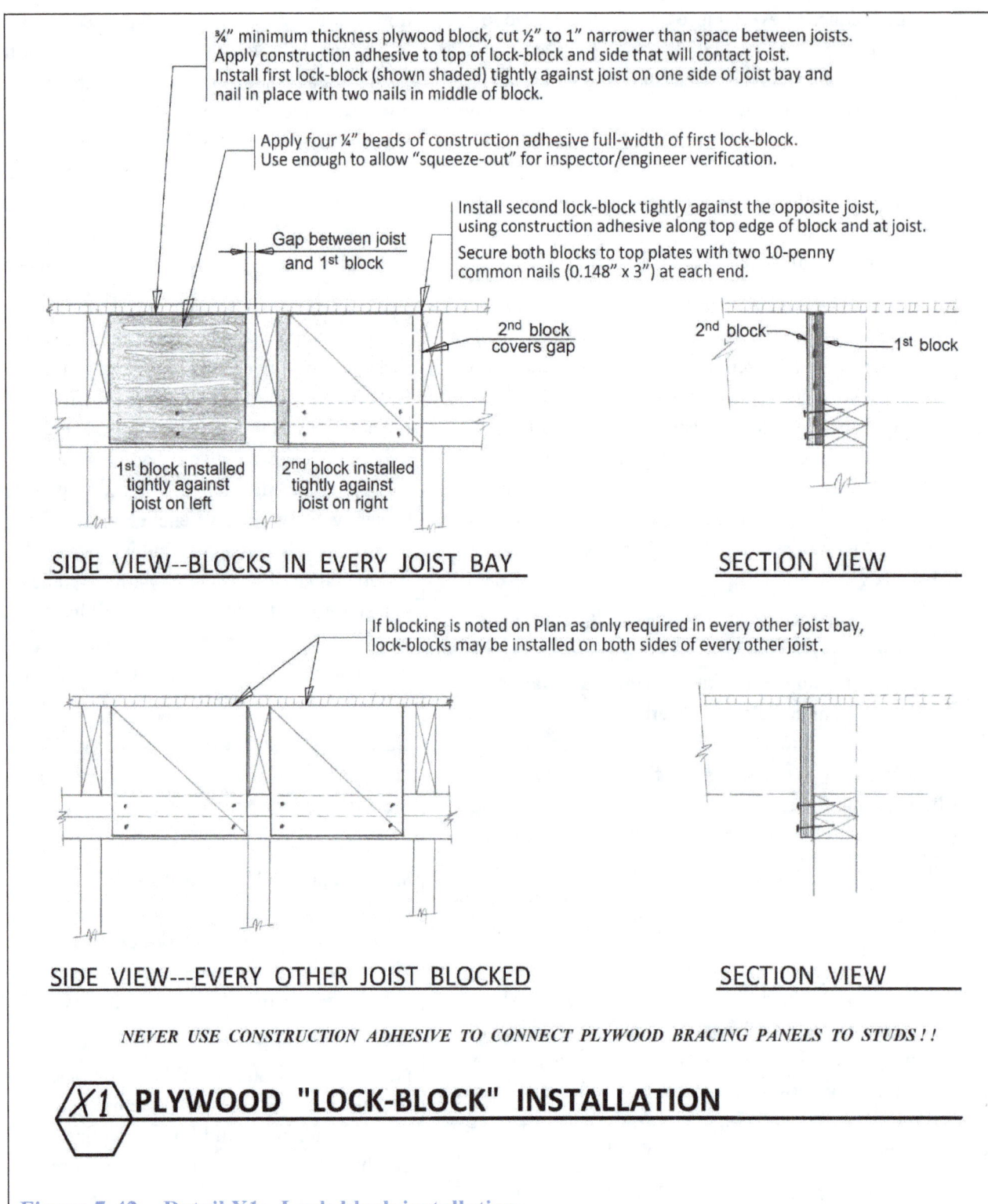

Figure 7-42 Detail X1: Lock-block installation

Detail of "Floating" Shear Wall for Hillside Construction

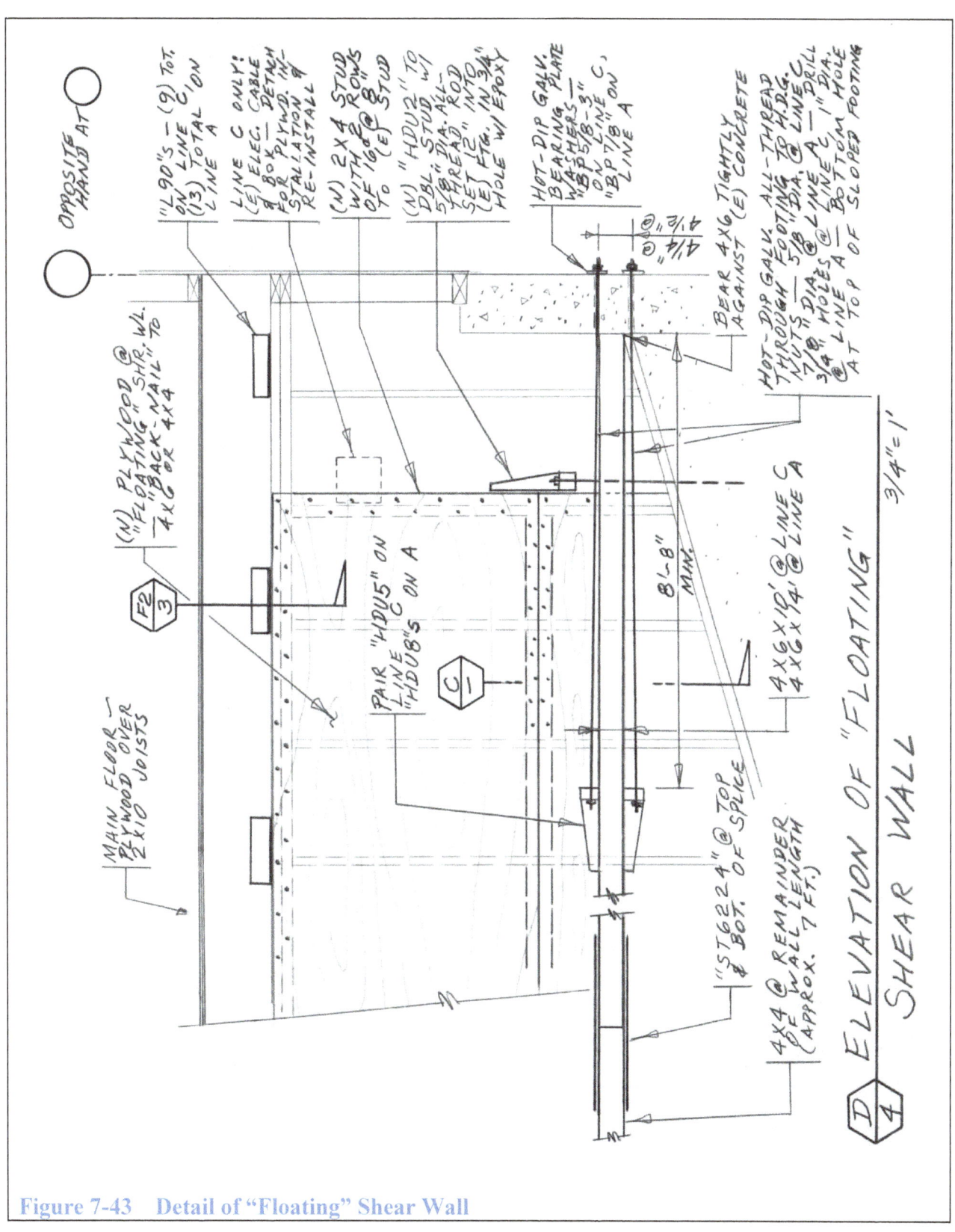

Figure 7-43 Detail of "Floating" Shear Wall

Section Detail for "Floating" Shear Wall

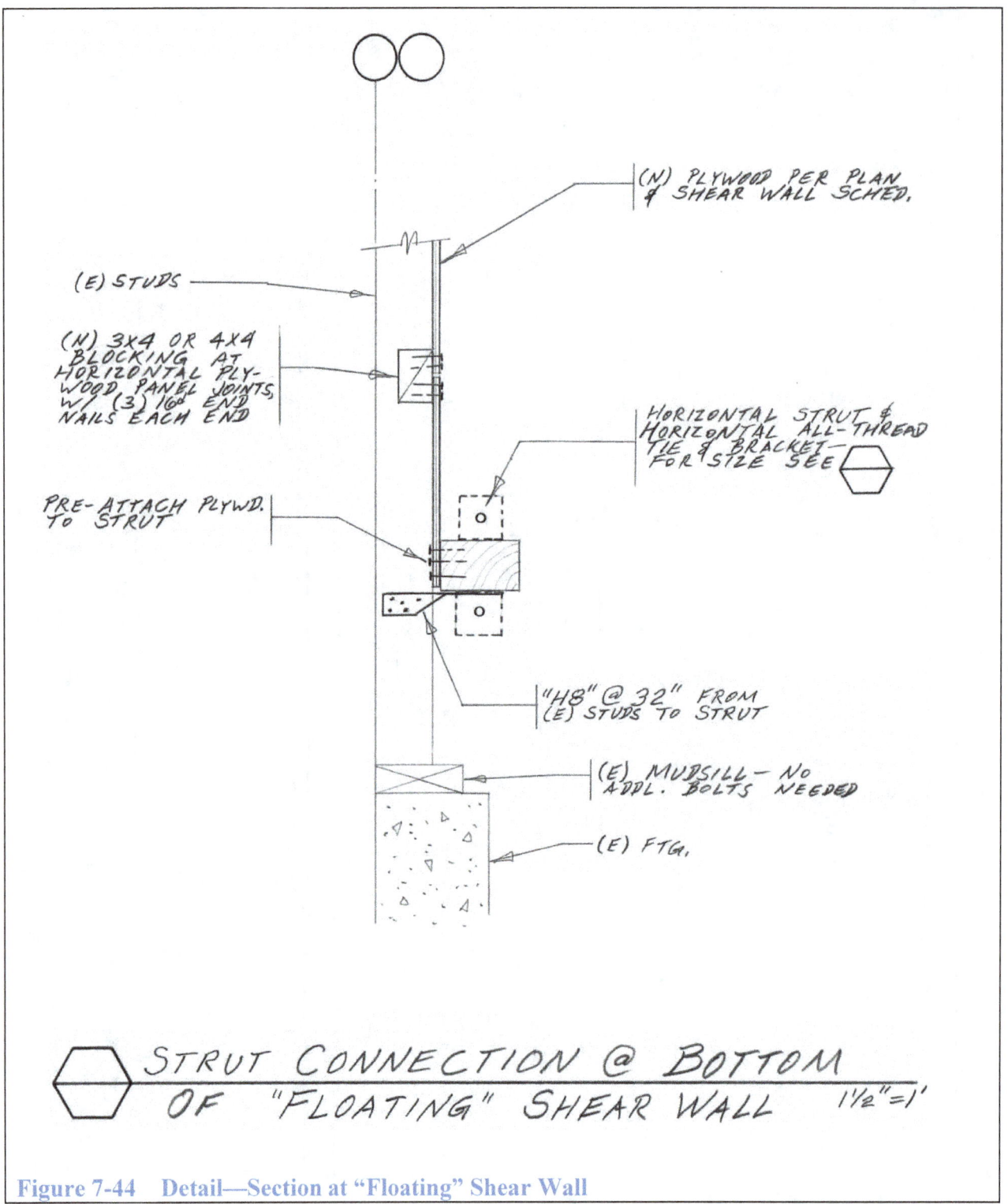

Figure 7-44   Detail—Section at "Floating" Shear Wall

Chapter 8 Retrofit Mistakes

This chapter shows some of the work that inspired this book—but not in a good way. Millions of dollars have been wasted installing hardware that will do little good to resist earthquakes. Seeing such work is very depressing but gives incentive to present better methods as outlined in earlier chapters.

Retrofit mistakes fall into four general categories: stock hardware mis-installed so it cannot effectively resist earthquake forces; "home-made" hardware devised without considering the nature of earthquake forces; and expensive installations that will do very little to resist earthquake. Sadly, the fourth category is just plain poor work quality. This is usually recognizable, but some common cases are addressed to preempt the argument that they are acceptable because "everybody else does it this way."

Another problem, related to building materials, is serious enough to deserve its own section: chemically treated lumber that causes severe corrosion in steel connectors (Section 8.5).

8.1 Mis-installation of stock hardware

The variety of mis-installations is staggering. Most mis-installations stem from the misunderstanding that earthquakes create only uplift forces. The following sampling of retrofit connections vary from amateurish to carefully executed installations, all with one thing in common: none of them are likely to provide resistance to earthquake forces acting from side to side.

8.1.1 Mudsill and related connections using sheet metal straps

Figure 8-1 shows some of the ways that a steel strap intended to resist tension forces is misused where we need to resist sliding forces. A tension connection between the mudsill, cripple wall, or floor framing and the foundation offers almost no protection against side-to-side earthquake forces. A retrofit bolt or UFP10 is needed here, along with plywood bracing panels.

Figure 8-2 shows additional examples of a very commonly misused hardware item. The manufacturer of these straps at one time showed them in a brochure demonstrating earthquake retrofit methods. This mis-installation has spread far and wide, even showing up on government-sponsored websites showing retrofit methods. One model of the strap shown really does have a rated capacity for lateral loads—but even when installed correctly it takes about seven of them to resist the same force as a single UFP10.

Figure 8-1   BAD: The manufacturer of these straps does not recommend using them to resist sliding forces that lead to most earthquake damage. Any of the retrofit anchors discussed in Section 6.2.1 would be better than these straps. For a complete retrofit, the cripple studs also need to have plywood bracing panels installed.
Photos: Left—Max Curtis, GGASHI; Right--author

| 1. BAD | 2. WORSE | 3. WORSER | 4. WORSTEST* |

Figure 8-2   *Photo 1*: Simpson "FJA" is rated for 185 pounds of horizontal force in line with the footing.   This connection is now considered obsolete and inferior to brackets such as the Simpson UFP10 (capacity of 1340 pounds).   Photo: Author
Photo 2: Simpson "FSA" (with longer 'tail' than the FJA) has no rated capacity at all to resist horizontal forces. The bolts into the floor joist are much too close to the bottom joist edge to achieve even the full uplift resistance.   Photo: Roger Drosd, GGASHI
Photo 3: Any hardware installed with a wood spacer between it and the concrete will have extremely poor performance. (Note that the installer doubled up the connectors because they looked weak.) Photo: John Fryer, GGASHI
Photo 4: Again, the FSA strap is rated only for tension, and then only when it is installed straight.   Photo: Paul Rude, GGASHI   * what's next—"Worcestershire?"

8.1.2 Hurricane ties used as Shear Transfer Ties

Figure 8-3 shows Simpson H2.5 "hurricane ties." These are great for resisting uplift—at least when all the nails are installed—but provide only 150 pounds lateral resistance (a Simpson "L90" uses the same number of nails, but has almost five times this capacity.) Many similar hurricane ties or "twist-straps" are used for this connection with equally poor performance. One of the few hurricane ties that is worth installing is Simpson's "H10A," when joist blocking is present; see Section 6.2.4.2.

8.1.3 Other hardware installations that must have looked like a good idea

The following photos illustrate many useless hardware installations. The variety is depressingly amazing. Figure 8-4 through Figure 8-14 show a sampling of mostly well-intentioned work that will provide almost no protection at all from earthquakes.

Figure 8-3   WEAK: This "hurricane tie" is not an efficient connector to transfer side-to-side forces from the floor framing to the top of the plywood bracing panel (in this case it is not even nailed to the joist).
Photo by Paul Barraza, GGASHI

Figure 8-4   BAD: This "T" strap does not have any load rating; even if it did, the plywood shim would degrade its performance—and it would tend to spin around the single bolt in the concrete.

Figure 8-5   BAD: Straight straps are bad for resisting side-to-side forces—bent straps are worse.
Photo: John McComas, GGASHI

Figure 8-6   BAD: Diagonal braces are not as effective as plywood bracing, and in any case must extend from top to bottom of walls. Perhaps these diagonal steel straps were chosen to fit around the existing pipes.
Photo: Paul Barraza, GGASHI

Figure 8-7   STRANGE: The tie-down brackets shown serve very little use anywhere besides the ends of plywood bracing panels. This is a very expensive installation of what serve only as mudsill anchors.
Photo: Paul Barraza, GGASHI

Figure 8-8   BAD: Diagonal bracing works in tension only (the saying "you can't push on a rope" applies).
Diagonal bracing is further limited by the connections at the ends of the braces—see close-up at right.

Figure 8-9   BAD: Close-up photo of bracing installation shown at left. Only **a few nails connect the straps to the original framing;** compared to plywood bracing panels this installation has very little strength.

Figure 8-10   BAD: This "FJA" strap is not even connected to the foundation— it holds only the stud from lifting off the mudsill. This wall needs plywood bracing panels and mudsill anchors.
Photo: Dave Heilig, GGASHI

Figure 8-11   BAD: The tie-down brackets shown would not effectively resist side-to-side loading.
Photo: Bill Londagin, GGASHI

Figure 8-12   WORSE: The threaded rods extending from the joist to connect this pair of tie-downs makes even uplift resistance a fantasy.
Photo: John Fryer, GGASHI

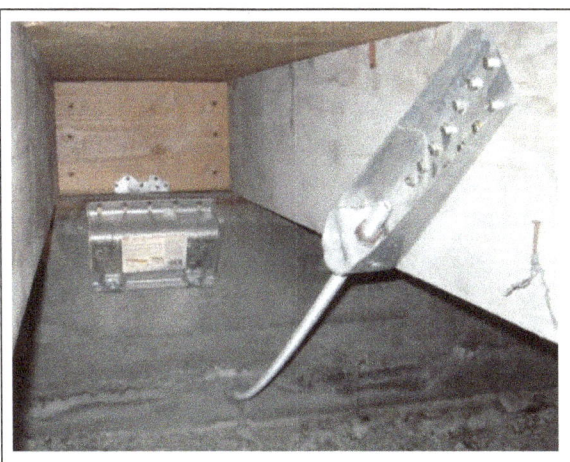 

Figure 8-13   Not Good: The tie-down bracket connecting to the joist on the right side of the photo will exert a substantial force downward on the joist during an earthquake. The slight curve in the all-thread rod running to the foundation will permit movement until it is drawn tight. This installation could be improved substantially.
Photo: John Fryer, GGASHI

Figure 8-14   BAD: This tie-down bracket serves no purpose unless plywood bracing panels are used. In any case, connection to the corner of the stud is bad practice; so is a connection using lag screws.
Photo: Paul Rude, GGASHI

8.2 *"Home-made" hardware and invented connections*

Custom earthquake retrofit installations may require engineered steel connections made to order by skilled workers in a steel fabrication shop. Some of the invented connections and hardware seen under houses did not attain this level of quality and design This section illustrates common problems seen with home-made items.

Only in the last 20 years did hardware manufacturers start making some of the connectors often needed in retrofit work. Before then (and in some cases since then), some well-intentioned connectors were developed in local garages or welding shops.

8.2.1 Angle-irons

The most common custom-made bracket in the SF Bay Area is the "angle-iron" connector as shown in Figure 8-15. The angle-iron connection was conceived to address the common problem where there is not sufficient space to drill down through the mudsill and into the foundation to install new mudsill anchors. It also has the supposed advantage of being able to resist forces in any direction—vertically, and horizontally from front-to-back or left-to-right. Resisting upward forces is usually not important, and the connection's capacity to resist horizontal forces is inefficient at best. FEMA recommends against this connection (see FEMA publication #547). There are several ways that the angle-iron connection can fail, which are discussed in detail in Appendix E

Some structural engineers recommend removing angle-irons, suggesting that they will cause damage during an earthquake. If you provide new connections such as UFP10s I do not believe you need to remove existing angle-irons. As of this writing, some other engineers still feel that there is a place for properly-designed angle-irons. Products like the Simpson UFP10 are designed to work in low-clearance situations, are easily recognizable, and have been tested to achieve certain loads.

Figure 8-15   FEMA recommends against using steel angles like the one shown above to resist earthquake forces in the direction of the foundation wall. This angle-iron is installed about as well as one could hope for. The two bolts in the concrete are spaced well apart; the top bolt is not too close to the top of the concrete, and the angle is installed tightly against the concrete and to the joist.
Photo: Paul Rude, GGASHI

Note that angle-irons are useful connectors in hillside homes (see Section 9.1.2) and possibly other situations, which are currently beyond the scope that I can address in this book.

Slight variations can have significant effects on an angle-iron bracket's connection capacity, even if they are installed perfectly. Each different configuration requires separate analysis. There are so many ways to install angle-irons badly that I fully agree with FEMA's recommendations against this type of connection. For any particular installation, the following factors can affect performance:

- Dimensions and thickness of steel angle used
- Distance between concrete anchors in footing (there MUST be at least two of them)
- Size of anchors connecting the angle iron to the footing
- Distance from top bolt in concrete to center of joist
- Placement of bolts in joist
- Joist size
- Presence of joist blocking
- Position of steel angle (must attach directly against concrete)

Figure 8-16 through Figure 8-23 show some examples of bad installations, along with explanations of their shortcomings.

Figure 8-16   BAD: These angle-irons have only one bolt connecting them to the foundation. Movement from side to side will simply pivot the angle-iron around the single bolt, twisting the joists lengthwise and splitting them; these are harmful and should be removed. Photo: Paul Rude, GGASHI

Figure 8-17   BAD: Single bolt to joist will worsen "cross-grain bending."
Potentially Bad: The distance between the two bolts in the concrete should exceed the angle length—otherwise lever action becomes a problem.
Photo: John Fryer, GGASHI

Figure 8-18 BAD: The wood spacer between the steel and the concrete has the structural equivalent of butter. This connection will act about the same as the one-bolt case shown in Figure 8-16.
Photo: John Fryer, GGASHI

Figure 8-19 BAD: The length of these angles is much greater than the distance between the two bolts in the concrete. The steel angle becomes a long lever that will easily exceed the concrete anchor capacity.
Photo: Paul Rude, GGASHI

Figure 8-20 BAD: The new wood blocks installed here have very minimal connections to the original structure. Connecting steel angles to the blocks provides very little strength.
Photo: Paul Rude, GGASHI

Figure 8-21 BAD: Even a small gap between the concrete and the steel will severely degrade the concrete anchors capacity.
Photo: Paul Rude, GGASHI

Figure 8-22 BAD: Nut was removed to expose a hole that was burned with a cutting torch instead of being neatly drilled; this would allow incresed movement.

Figure 8-23 BAD: Just completely bad in every way possible; connection will pivot in two directions, bend, and split the joist.
Photo by David Venable, GGASHI

8.2.2 Miscellaneous custom hardware

The variety of custom hardware found under old houses is testament to the unusual conditions that contractors face when trying to install retrofit work. The following photos show a sampling of installations. Some of these are purely one-of-a-kind, while others are less rare.

Figure 8-24   BAD: Neither of these steel straps will resist appreciable side-to-side force. The bent strap on the left would not even resist uplift forces until it straighened out. Photo by John Fryer, GGASHI

Figure 8-25   BAD: Only one bolt connects this steel plate to the concrete, which will allow the plate to pivot around the bolt. Photo by John Fryer, GGASHI

Figure 8-26   BAD: The welded angle has the same problem as the assembly shown in Figure 8-17, plus a few additional problems. Photo by Paul Barraza, GGASHI

Figure 8-27 BAD: Installations using scrap materials are disturbingly common. Photo: Gary Tucker, GGASHI

Figure 8-28   BAD: Even though very heavy-duty, these bent plates would pivot around the single concrete anchor and have no connection to the mudsill. If the mudsill did tend to lift up off of the foundation, these brackets would cause cross-grain bending that would split it lengthwise. Photos: Brian Cogley, GGASHI

8.3 Expensive retrofit work that provides relatively little protection

Crawlspaces are really inconvenient places to work. Designers should carefully weigh the expense and the effectiveness of proposed connections. The following sections present some common cases of very poor use of retrofit dollars.

8.3.1 Connecting many short cripple-posts to the beams they support

Figure 8-29 shows a retrofit connection securing an interior post to a floor beam. Some houses have dozens of such posts supporting the first floor system.

NOTE: There is a huge difference between interior posts in a crawlspace with a perimeter foundation system, and a post-and-pier foundation (which is essentially not a foundation). Bracing posts in the former case is hardly ever needed; in the latter case, a new foundation is usually needed.

Post-to-beam connections are an enormous waste of money because they are difficult to install but do essentially nothing to keep a house from moving back and forth during an earthquake. The nature of the connection cannot resist appreciable force. In an otherwise properly retrofitted house, the only force that a post connection would need to resist is the force generated by the post itself—a few pounds at most.

The photos in Figure 8-30 are used in FEMA training materials to indicate the importance of secure post connections. What these photos illustrate to me is that *the post connections did not fail,* **even though the building shifted a foot or more.** If the posts remained connected at top and bottom while the building moved that much, think how well they would have performed if the cripple walls of the buildings had been braced and kept them from moving! The deficiency here was not the post connections—it was the weak cripple walls that allowed the building to move in the first place.

Figure 8-29 **Wasted money;** interior post-to-beam connections almost never need supplementation Photo: Paul Rude, GGASHI

Figure 8-30 These posts stayed connected even though the buildings shifted a foot or more. Lesson: Brace the perimeter of the building and interior posts will be fine. FEMA photos.

If the posts were originally nailed to the beams, in my opinion additional connections are not warranted. (For a case when you should worry more about a secure post-to-beam connection, see Figure 8-31.) If the floor system moves enough that post-to-beam connections are in danger, it simply means you should have spent more effort installing connections around the perimeter of the house.

Before you spend much time supplementing post-to-beam connections, ask the following:

1. Could you remove one or more posts without causing any portion of the building collapse? (Note that "*collapse*" is very different from "*sag*.")
2. Is the access to the posts too low to allow crawling on your hands and knees?
3. Can you see any sort of existing connection between the post and beam, such as toe-nails?

If you answer "yes" to any of the above, supplemental connections are probably not needed. Usually the more work it would take to install retrofit connections, the less likely you need them. If access to the posts is very difficult and there are many of them, upgrading the connections is likely unnecessary.

If you have only a few posts and you can get to them easily then it may be worth the trouble to connect them better.

Report from a Home Inspector after the 1989 Loma Prieta quake:

"The house was very close to the epicenter. One end was single-story and built with 2x6 plank flooring over 4x6 beams. The perimeter of the house was built right on the foundation, with no cripple walls. The sills were bolted to the footing and survived just fine."

"All but two of the interior posts supporting the 4x6 beams were originally toe-nailed in place. The two posts without any toe-nails were in the center of the crawlspace. They ended up slightly out of plumb after the quake, i.e., they were resting on one corner at the bottom and the opposite corner at the top. Since their height was now determined by the diagonal length between opposite corners, the floor was slightly crowned over them."

"**The floor did not move sideways—the posts were still in their original position horizontally with respect to the pier blocks below them; all that was different was that the posts were no longer plumb.**"
— Douglas Hansen, *Code Check*

Figure 8-31 The room extension shown is supported on a "post and pier" foundation, rather than a perimeter foundation. Losing support from this post would cause collapse, and strengthening the connection is reasonably easy. Even so, supplementing the connection will not provide any bracing capacity—that would require a new foundation or other work under the main portion of the house.
Photo by Michael Brady, GGASHI

Figure 8-32 shows what you find under hundreds of thousands of houses built since the 1950's or so. You could temporarily remove every other post and not suffer much damage. The short posts would be very unlikely to move at all during an earthquake anyway. Access to the posts varies from annoying to hellish. For this house I would recommend no more than installing UFP10s and supplementary connections from perimeter framing to the mudsill.

Crawling on your belly makes distances seem longer. Even unobstructed areas are hard to access. Leave these posts just as they are.	Crawling is one thing—now consider pulling an extension cord, work-light, compressed air hose, and tool tray behind you.
	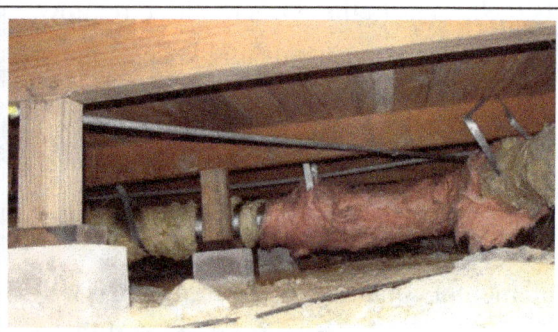
Previous workers scraped all the insulation off this duct. Duct joints can get knocked apart, allowing contamination of the duct system.	"You can't get there from here." Anyone specifying post connections in such places should be required to go inspect each one.

Figure 8-32   The photos above show a crawlspace under a house built in the 1960's. Similar construction dominated the housing industry in that era. Supplementing post-to-beam connections in such houses is extremely unproductive; perimeter connections from foundation to framing above are usually all that is necessary in this sort of house (assuming other hazards such as split-level construction are not present).

Plywood gussets such as the ones shown in Figure 8-33 and Figure 8-34 were, for a time, illustrated in a guide by the American Plywood Association on how to strengthen against earthquakes. In September, 2012, I spoke with the engineer in charge of the APA's technical publications and asked him about their gusset recommendation. A couple of days later he called to say that he had reviewed the document in question and the APA had decided to cancel its further publication.

Figure 8-33 This giant gusset is beyond useless: it obstructs use of the garage. Note the plywood in the background; if installed correctly, it makes the gusset totally unnecessary.
Photo: Paul Barraza

Figure 8-34   Wasted effort—Plywood gussets offer almost no earthquake resistance. APA—the Engineered Wood Association used to suggest these plywood gussets. In 2012 they canceled publication of the document that recommended connections like those shown above. Connections from the posts to the concrete are also far from necessary.

By their very nature, post-to-beam connections cannot resist much side-to-side force. Bracing the perimeter cripple walls of a house provides earthquake resistance much more efficiently than bracing the posts. Again: Any connection strength beyond that needed to resist a force equal to the weight of the post is an unnecessary waste of money. Figure 8-35 shows a very expensive yet useless installation.

In my opinion, supplementing post-to-beam connections is probably the most over-hyped part of earthquake retrofit lore. Post-to-beam connections are mentioned on contractor's websites, in blogs, newspaper articles, FEMA documents, home-inspectors reports, etc. Strengthening post-to-beam connections is the least economical retrofit strategy in most houses.

Figure 8-35   Wasted effort--One can only hope that as much effort was spent bracing cripple walls as was spent on these post connections.
Photo: Paul Barraza, GGASHI

Figure 8-36   Some post-to-beam connections really do need supplementation. (This post was added to reduce floor sag, so its loss would not be catastrophic.)
Photo: Michael Brady, GGASHI

8.3.2 Post connections to piers

Figure 8-37 shows some methods used to supplement the connections from cripple posts to their supporting piers. Uplift is not a concern at this connection. The work shown represents anywhere from $50 to $100 worth of wasted retrofit effort; multiply by ten or twenty posts.

Figure 8-37   **Wasted effort & possibly harmful. Original post base connections hardly ever fail in earthquakes. Brace the perimeter walls and the posts will not go anywhere. Photos: Left, Center—Paul Barraza, GGASHI; Right—author**

If we consider possible failure modes and consequences, we find that anchoring the post to the pier could be worse than letting it tap-dance on the pier block in the unlikely event there was that much upward force. Piers such as those shown often extend only 4 inches below the ground surface. Such an installation would resist only a couple of hundred pounds of uplift, and then the pier would lift up out of the ground. Once the pier lifts up it would be next to impossible for it to fit perfectly back into place. Soil would cave into the hole under the pier, which would result in the post being too high after a quake. Now you need to crawl under the house and remove any pier that was pulled up out of the soil, clean out the hole, and pour a new pier. If you had left the post-to-pier connections alone, maybe the post would have pulled up off the pier--no big deal, you just reattach it. It won't have gone far. One can come up with hypothetical situations where tie-downs would be beneficial in the cases shown, but it seems extremely unlikely—and hardly worth the investment.

8.3.3 Connections to underside of floor diaphragm

Figure 8-38 shows a sheet metal connector attached from the underside of the floor sheathing to the face of retrofit plywood bracing panels. Since one does not want to drive nails up through the floor and into the owner's sleeping cat, these connectors must be fastened with short screws. Assuming that the flooring is only ¾" thick, you must use very short screws. Short screws have poor holding ability so you have to use more of them. Requiring a worker to hold onto short screws and drive them into place overhead while lying in a cramped crawlspace probably violates the Geneva Convention.

This connection is almost never necessary; see Section 4.1.2.2. The International Existing Building Code assumes that existing connections are adequate at the perimeter of floor diaphragms. Considering that this is a very difficult connection to install, when I need to connect new members to the underside of existing floor sheathing I specify structural adhesives.

Figure 8-38 Difficult, inefficient, and almost always unnecessary; there are better alternatives to these sheet-metal connectors screwed to the underside of the floor.

8.3.4 Costly FEMA detail

FEMA 547 (see "References" section) suggests the connection shown in Figure 8-39 and states that *"Where cripple walls are 14" tall or less, wood structural panel sheathing may no longer provide reliable bracing of the studs, and splitting of the studs becomes a more significant concern. For this configuration, use of solid blocking between the studs is recommended instead of sheathing."* This is confusing; it suggests splitting of the studs may be a separate issue from the "unreliability" of sheathing on short cripple walls.

I contacted FEMA about this and other concerns. Their response noted only that contractors had reported problems with short cripple studs splitting. Had the contractors been asked to suggest solutions, I expect that

Figure 8-39 Difficult to install: FEMA suggests this very labor-intensive installation to avoid splitting the stubby cripple studs with plywood nails. Installing plywood and pre-drilling a dozen nail holes to prevent splitting would be much easier.
Photo: Matt Cantor, GGASHI

they would have proposed pre-drilling holes in the short studs, or installing new 4x4 posts. Either of these methods would effectively prevent splitting. The blocks shown required nine carefully cut pieces, whereas one of the suggested alternatives would require at most two posts, three mudsill blocks, and a piece of plywood—and only the posts would need precise cutting.

Other concerns with FEMA's suggested connection include the greater rigidity when compared to plywood bracing panels, potential lack of ductility, and absence of test results to support its use. The block installation shown has no connection from the mudsill to the top plates (FEMA is silent on how to attach the blocks) and almost certainly does not provide the same sort of ductility that a plywood shear wall does. Furthermore, most engineers would require connecting all four sides of a block—which would probably split the studs where the blocks are nailed to them, thus defeating the stated purpose of this detail.

For new construction, the International Residential Code offers a substitute for solid blocking. IRC Section R602.9 allows using plywood panels nailed to the top cripple wall top plate and to the mudsill as an acceptable substitute for solid blocking. Such an installation requires that the mudsill width equals the stud widths, so we need to address this before we can attach plywood.

The installation shown in Figure 8-40 might be a contractor's solution to the problem of split cripple studs, which eliminates splitting much more efficiently. Instead of a cluster of blocks, a single length of 2x12 connects to the top plate of the short cripple wall behind. The bottom of the 2x12 is connected with framing connectors to the original mudsill. (Since the mudsill is not wide enough to accommodate the 2x12 and the outstanding legs of the connectors, the connectors were nailed to the mudsill before installing the 2x12. The leg of the connector attached to the mudsill points away from the crawlspace, and is covered by the 2x12. Then the connectors were nailed to the face of the 2x12.)

Figure 8-40   This alternative avoids nailing to short studs with much easier installation than the blocks shown in Figure 8-39. The "L90" framing connectors were installed first; the 2x12 was then installed and conceals the horizontal leg of the L90s.
Photo: Brian Cogley, GGASHI

8.4 Inferior construction methods

Some construction techniques took hold decades ago; just because a construction method has been used for a long time does not make it effective. In some cases new tools and hardware make older, once acceptable methods obsolete. Contractors, home inspectors, and real estate agents have likely seen these methods more than once.

Figure 8-41 and Figure 8-42 show a few commonly-found practices in plywood installation. One misconception about plywood bracing panels seems to stem from the idea that if *most* of the wall is covered, then all is good. We are not building a fence to keep tumbleweeds out— all edges of all panels must connect to framing members. Likewise, even though they do not take much material out of the plywood, the over-cuts at the flexible conduit weaken the panel.

Figure 8-41   BAD: *Left*: Covering most of the wall with plywood is not enough; *Center*: An honest bad installation; no effort was made to hide the slot for the pipe; *Right*: Slots with sharp inside corners concentrate stresses; "over-cuts" that extend beyond the corners of this cutout for electrical cables are even worse. Best practice is to use a hole-saw to start cut-outs so they have rounded corners.
Photos: Center—Paul Rude, GGASHI; others—Author

Figure 8-42   BAD: *Left*: Slots cut into panels so they can fit around existing utilities must be patched. See Section 6.4.2; *Center*: Drywall screws are brittle and can snap off during an earthquake; use only nails to attach plywood. *Right*: Uneven nailing makes plywood bracing perform worse. The blue lines show locations of nails along the edges of this panel; for the same panel edge length, there are 11 nails on the left side and seven on the right side.   Photos: Center: Paul Barraza, GGASHI; others--Author

The photos in Figure 8-43 and Figure 8-44 show examples of unacceptable work. Sadly it is easier to get away with shoddy work when it is hidden away in a crawlspace. Plumbers and other trades workers are legendary for destroying structural components.

  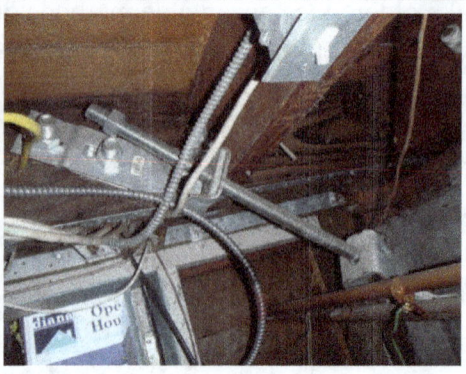

Figure 8-43 **JUST PLAIN BAD:**
Left: Severely leaning bolt. Photo: Paul Barraza, GGASHI
Center: Mudsill held down by wood scrap under a mudsill anchor.
Photo: Paul Rude, GGASHI
Right: I don't know what that pair of hold-down brackets was supposed to do, but the bracket at lower right of the photo resting on top of the pipes is probably not doing it.
Photo: Gary Tucker, GGASHI

Figure 8-44 **JUST PLAIN BAD:** *Left*: Plywood hacked out by a plumber. This slot could not be fully patched with glued-on plywood because the electrical outlet obstructs ample overlap for gluing; the water pipe will also keep a patch from lying flat against the original plywood.
Center: Plywood nails driven from the far side of the wall that did not hit the framing members (and when they did, as with the stud on the left side of the photo, the studs split);
Right: Plywood panel joints that do not occur on framing members (so plywood edges cannot be nailed to anything), and poor connections as well.
Photos: Left, Center: Paul Barraza; Right: Max Curtis (all GGASHI)

8.5 Pressure-treated lumber: a ticking time-bomb since... 2004?

Alert: This section is *long*; there is a great deal of misinformation to address, and the subject involves some technical information that needs explanation.

Imagine constructing a house using building materials that will quietly destroy each other. Millions of homes have been built using such materials; the hardware pictured in Figure 8-45 shows severe corrosion of steel connectors in contact with chemically treated lumber. Rust weakens connectors, making them more likely to fail during an earthquake. Future failures of structural connections will surprise a lot of building professionals and homeowners; if you have this material in your home you still have time to avert disaster.

Figure 8-45 Most wood now sold to resist decay also attacks steel connectors. The red circle shows a rusted concrete anchor. At center is a severely corroded H10A (compare to Figure 6-34). This installation was about 10 years old when photographed. Photo: Matt Cantor, GGASHI

8.5.1 Background information

Ever since the first building code there has been a requirement for mudsills or other wood members in contact with earth or footings to be decay-resistant. The codes allow foundation-grade redwood or cypress, or preservative-treated wood. "*PT lumber*", or even just "*PT*" is the term used for wood that has been made artificially decay-resistant by the addition of preservatives. Technically the proper term is "*preservative pressure-treated*" lumber, as wood can also be pressure-treated with chemicals to make it fire-resistant (and possibly for other traits). I will stick with the term PT lumber here.

8.5.2 Alphabet soup—and unintended consequences

The chemicals: Until 2004, the most common treatment chemical used in PT lumber intended for residential construction was chromated copper arsenate, or "CCA." You may find PT lumber stamped with "CCA" or "CCA-C," (or some other suffix letter; the trailing "C" is the third formulation they came up with, after A and B). Arsenate is a chemical compound that includes arsenic, which has been a known human poison for centuries. The chromate is probably not good for us either. Thankfully the CCA binds with the lumber and the toxins pretty much stay in the wood.

Since 2004, CCA has not been allowed as a treatment for wood used in residential construction in the US. Some of the more common new chemicals used for preservatives include CA (Copper Azole), ACZA (Ammoniacal Copper Zinc Arsenate), CC (Copper Citrate), ACQ (Ammoniacal Copper Quaternary). All of the preceding chemicals bond with the wood cells, giving a "waterproof" treatment suitable for fence-posts, decks, etc. Figure 8-46 shows several labels from PT lumber.

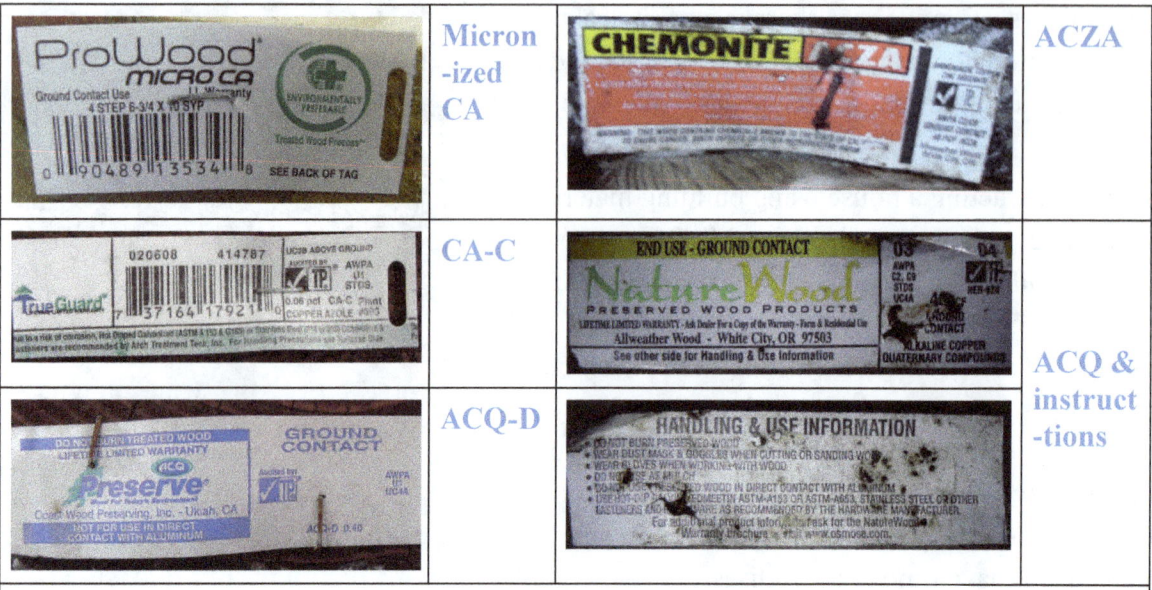

Figure 8-46 Tags from several varieties of pressure-treated lumber found on jobsites; all of them are rated "ground-contact," which has higher chemical content than the minimum needed to protect lumber used as mudsills.

All of the preceding treatment compounds contain copper. Copper and steel, in the presence of water and oxygen, create a "galvanic reaction" (*galvanic* and *"galvanized"* both give tribute to Italian scientist Luigi Galvani). This reaction occurs even with very small amounts of water, such as from humid air in a crawlspace. Oxygen molecules will travel through wood, so embedding nails or other fasteners into wood does not protect them. The steel corrodes as a result of this reaction. Corrosion is the chemical term for "rust." Figure 8-47 shows some very disturbing examples of rapid corrosion.

The "white rust" often seen on galvanized steel connectors is the first corrosion by-product: zinc oxide. After

Why the "teeth-marks" in PT lumber?

To speed up the process of getting the preservatives into the wood, the wood is immersed in chemicals inside huge vessels that are then pressurized to force the mixture into the wood cells.

Some species of wood will accept the chemical treatments more easily than others; wood that is difficult to treat otherwise is usually "incised" by sending the lumber through rollers studded with miniature knife-blades. Douglas fir is difficult to treat and needs to be incised for most chemicals. The "hem-fir" species group is also easier to treat, but woods in this group are softer and do not provide the same connection capacity as Douglas fir. (Since mudsills are not high-demand members, the reduced capacity can be addressed easily by using more edge-nails when connecting bracing panels.) Southern yellow pine accepts treatment readily and does not require incising.

the zinc corrodes away, the underlying steel is no longer protected and it begins to rust. **This process continues until all the steel has corroded away. ALL of it.** The time it takes depends on many variables—so: do you feel lucky? Note that hardware manufacturers advise if you see *any* "red rust" on a connector then it should be replaced.

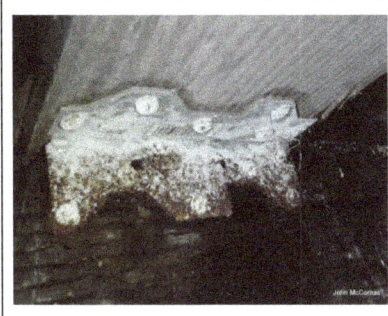  

Figure 8-47 Photos show severe corrosion that occurred within five years of construction.
Left: "White rust" turning to red rust. Photo: John McComas, GGASHI
Center: Remains of nails used to fasten PT plywood. Photo: Keith Tarkington, GGASHI
Right: Mudsill anchor photographed by author in May, 2014.

Compare the photos above with the anchor bolts shown in Figure 8-48, which were installed in either CCA or redwood mudsill material. The threads on the anchors are still well-defined, whereas the threads shown in Figure 8-47 are rusted badly enough that removing the nut would be impossible.

Figure 8-48 Bolts shown above were installed before PT chemical reformulation and show minimal corrosion.
Left: In a CCA mudsill, approx. 35 years old; Right: In a redwood mudsill, 80 years old.

The Chemistry of Corrosion

The reaction between different metals is an electro-chemical process where electrons flow from one of the metals toward the other one. Chemists refer to one metal (the one that corrodes) as the "anode" and the other as the "cathode."

Any particular metal will act as either the anode or the cathode depending on what other metal is present. Chemists rank metals on the "Galvanic scale." Metals that corrode easily (magnesium, zinc, aluminum) are at one end of the scale. Metals toward the other end of the scale include gold and platinum. Metals that do not corrode easily are called "noble" metals. You can find pure gold nuggets in nature because gold lasts millions of years without corroding. Metals that corrode easily are almost impossible to find in pure form in rock formations—they exist almost exclusively as oxides.

Iron (the main ingredient in steel) is near the middle of the Galvanic scale; if it is paired with zinc, it remains stable. When paired with copper, it corrodes.

(continued next page)

How much copper is in the wood?

The American Wood Protection Association (AWPA) generates specifications for various "use categories" for lumber and minimum treatment levels. Most PT lumber used in construction comes in one of two "retention levels" (the amount of chemical retained in the wood after the treatment process). The lowest level is "above ground," followed by "ground contact." Higher retention levels are available for marine use or wood pilings.

The amount of preservative remaining in the wood is measured in pounds of chemical per cubic foot of lumber. The amount required depends on the type of chemical, since some are more effective at protecting against decay. For ACQ and CA to meet the minimum "above ground" use category, each cubic foot of treated wood must contain 0.25 pounds of chemical. For "ground contact" the minimum retention level is 0.40 pounds of chemical per cubic foot of wood.

8.5.3 Most suppliers stock only the most corrosive wood

People use PT lumber for mudsills, fences, planters, retaining walls, play sets, decks and so forth. Some of these uses require the "ground contact" level of chemical treatment. Lumber suppliers have apparently gotten blamed when a contractor or homeowner installed "above-ground" rated wood in such cases, and chose to stock only the "best" grade of lumber. Between the desire to avoid confusion and save space that would be needed to stock two different sorts of lumber in many sizes and lengths, most suppliers carry only lumber treated with higher preservative levels. Unfortunately this means the PT lumber you buy for a mudsill

---

The Chemistry of Corrosion
(continued from previous page)

You may wonder why steel rusts when it is not in contact with another metal. There is enough physical difference in the steel that small regions will act as anodes and other regions act as cathodes. In the photo below, simply bending the reinforcing bars changed the steel properties enough that the bent regions of the bars became anodic relative to the straight sections, causing them to rust first.

When the initial anodic regions rust away, other regions are next in line. Thus if we have metal items that we want to keep from corroding, we need to protect them somehow. One way to protect them is to coat them with paint to keep the moisture and oxygen away. Another method is *cathodic protection,* i.e., protecting the cathode from corrosion.

Cathodic protection uses "sacrificial anodes." The sacrificial anodes give their lives to save the more noble metal you want to protect. Boat owners may be aware of zinc anodes used to protect their steel boat hulls. *(continued next page)*

has 60% more copper in it than necessary. More copper means faster corrosion of steel fasteners.

8.5.4 Bolts versus nails

Most bolts shown in the accompanying photos probably still have almost their full strength. I am much more concerned about nails driven into PT lumber. The nails are much smaller than bolts and would corrode faster. I would photograph nails if I could—in most cases they are covered with siding or stucco. You will never know the nails have failed until sheets of plywood or stucco peel away from the mudsill during an earthquake.

8.5.5 One solution: use Stainless Steel hardware and fasteners

Stainless steel contains chromium and nickel, which make it more corrosion resistant than copper. You may be stuck with nasty PT lumber that someone else installed when they replaced your foundation. In this case the only way to assure long-term performance of connections to the treated lumber is to use stainless steel fasteners and connectors. *Every* metal component that connects to, or passes through the treated lumber should be stainless steel. Stainless steel nails should be available at most local lumber yards, but you may have to special-order other components. The Maze Nail Company makes stainless steel nails. GRK Fasteners makes self-drilling structural screws. Note that Simpson Strong-Tie and other companies offer some of their connectors in stainless steel, but the cost of stainless is about *ten times* that of their regular connector line. Clearly you want to avoid the need for stainless steel fasteners.

> **The Chemistry of Corrosion**
> *(continued from previous page)*
>
> Aluminum or zinc anodes are commonly used to protect steel. Magnesium anodes are used to protect aluminum (magnesium could also protect steel, or any other metal more noble than it is; but magnesium is more expensive than aluminum).
>
> Galvanic corrosion occurs because our metal parts create a crude battery. Since the process involves very weak electrical current, the anode (metal that corrodes) does not have to be in contact with the cathode—just joined to it by a wire or other metal part. The photo at left below shows a concrete anchor with significant corrosion just three months after installation, and it was not even touching the ACQ-treated lumber; it was connected to a UFP10, which was in turn attached to the ACQ lumber. Most importantly, *the process does not stop until all of the anode material has corroded.* Sacrificial anodes must be replaced periodically, or protection is lost for the boat hull, oil pipeline, or structural framing connector. The photo below right shows a sidewalk access cover for Pacific Gas & Electric Company to replace an anode that is protecting an underground pipeline from corroding; if they did not replace the anodes regularly, the pipelines would corrode and eventually leak.
>
>

8.5.6 Addressing the problem, or ignoring it?

The wood-treating industry and the hardware manufacturers really don't want to tell contractors that durable fasteners and connectors will cost ten times as much as what they were accustomed to paying before the switch to new PT formulations. The industries use a few ways to avoid telling you straight out that you should use stainless steel connectors with treated wood.

Tactic #1: Misdirection & denial: One of the early "answers" from industry was to deny that there was a problem. The following is a quote from a preservative industry document dated 2004 (still at large on the internet as of September, 2014) regarding corrosion:

> *"It is generally recognized that the potential for fastener corrosion in forest products based building materials used in an interior exposure environment is minimal because the equilibrium moisture content of the wood is maintained at a level that does not support corrosion reactions."*

The above quote is generally recognized as the sort of statement intended to be as confusing, vague, and misleading as possible. What they mean is, "it is generally recognized that lumber used indoors will probably be too dry to allow steel to rust." The most misleading term is *"it is generally recognized."* Who recognizes this? And based on what? Testing? Hope? Unrelated studies? The document asserting this "fact" quotes a study of moisture levels in wood, dated 1988. When we are talking about corrosion caused by chemicals that were not even *invented* in 1988, this reference by someone who wants to sell me their product seems unhelpful at best. The second-most misleading term is *"interior exposure environment."* Does this mean inside the heated or air-conditioned part

Figure 8-49 This installation was four years old when photographed. Note the red circle at the top of the picture, which shows a rusted nail that is just barely touching the PT ledger.
Also note that the PT ledger is separated from a concrete foundation wall with tarpaper, which slightly reduces the moisture that would wick into the wood from the concrete.
The manufacturer of the joist hanger shown recommends replacing hardware that has any "red rust" on it. Hint: Replace with stainless steel if you do not want to repeat this process every four years. Photo by John Fryer, GGASHI

of your home or the damp crawlspace beneath it? One reason that PT lumber is used in contact with concrete is that the concrete wicks moisture up out of the ground, resulting in elevated moisture levels in the mudsill. Considering the severe corrosion of steel connectors seen in the last few years, as shown in Figure 8-50 through Figure 8-53, I am not reassured by this pitch from the wood-treating industry.

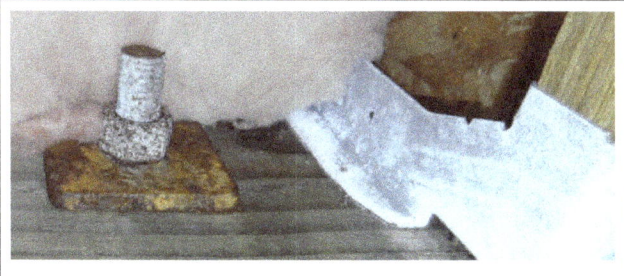

Figure 8-50 Six-year old installation in a mechanically vented crawlspace, where an exhaust fan was installed to provide reliable ventilation. It would take a fairly long time—maybe decades—for this bolt to corrode to the point it lost significant strength. Nails are a different matter; see below.

Figure 8-51 Rusty edge nailing of a shear panel, approximately five years after construction. Contractors have reported almost total corrosion of nails in just a few years. Photo by Skip Walker, Silicon Valley ASHI/CREIA

Tactic #2: Minimize the problem: After the new treatment chemicals had been corroding fasteners in the real world for a few years, manufacturers started including warnings in the fine print on their labels. One suggestion is that regular inspection and maintenance of the hardware will allow you to use the less-expensive fasteners they sell. "Regular inspection and maintenance" is possible—but, let's face it, highly unlikely—on a deck or a fence. For most connections that are crucial to your house's integrity, inspection & maintenance is completely impossible: how do you inspect or maintain the nails driven into a PT mudsill to attach the bottom edge of the exterior sheathing on your house if the nails are covered by siding or stucco?

Figure 8-52 Accidental experiment—the builder left a couple of foundation anchors lying on the dirt under a house (photo above left). Photo above right compares one of them to an anchor installed in PT lumber nearby. The anchor that had been left on the bare ground still has shiny portions, whereas rust completely covers the anchor installed in the PT lumber.

Tactic #3: Pass the buck: Fine print on PT lumber labels and hardware manufacturer's brochures tosses all the responsibility for corrosion to the consumer, just saying to make sure you select the appropriate fasteners to use with the chemicals in question. One hardware manufacturer states that their products perform well under controlled tests, but also notes that the test results may not apply to actual field conditions and suggest using stainless steel if you are unsure about the specific installation. If *they* are not sure, how can *you* be sure? Unless you have a degree in metallurgy and a laboratory to do your own tests, the hardware manufacturers are essentially recommending stainless steel.

For years, Simpson Strong-Tie's catalogs have listed a "corrosion exposure" level of Low, Medium, or High. Their 2015-16 catalog adds a new level: *severe*. Their catalog indicates what fastener materials are recommended for connecting to treated lumber, based on the treatment level, type of chemical, and whether the installation will be dry or wet. For *any* of the copper-containing wood treatments at

---

All Galvanizing is not Equal

"Galvanizing" generally means to coat a metallic item with zinc. The zinc is applied in one of three ways: electroplating, mechanically, or by dipping the item into a bath of molten zinc. Each method has advantages and disadvantages.

Electroplating uses an electric current to deposit a metal (in this case, zinc) onto the base metal being plated. The plating is deposited one atom at a time from a bath containing a solution of a zinc salt. Electroplating deposits a fairly uniform layer of zinc, and gives parts a somewhat shiny, silvery appearance. This method is also called "electro-galvanizing." For something as crucial as an earthquake retrofit system, electroplating does not deposit enough zinc onto the base metal to offer adequate protection from corrosion. That leaves mechanical or hot-dip galvanizing.

Mechanical galvanizing is done by tumbling a bunch of metal parts in a rotating drum with powdered zinc. Enough zinc is pounded onto the surface of the metal to offer protection. This method is used to coat threaded parts that might otherwise turn to blobs if they were dipped into molten zinc. This method is sometimes referred to as "tumbling." Mechanically galvanized parts have a dull-gray finish. Simpson "SDS" screws are currently supplied with mechanical galvanizing.

Hot-dip galvanized ("HDG" for short) parts are dipped into a kettle of 900-degree molten zinc. Many galvanizers have kettles big enough to dip 60-foot street-light poles. The molten zinc heats the base steel to the point where four different iron-zinc alloy layers are formed. Two of these alloy layers are actually harder than plain steel. HDG material has a spangled appearance. Most sheet-metal framing connectors are hot-dip galvanized.

Hot-dip galvanizing can coat the entire surface of a part with zinc, which offers the best protection from corrosion—however, see the sidebar on the following page for some bad news.

"higher chemical content" levels they recommend stainless steel. They define "higher chemical content" as the levels required for wood to be rated as "ground-contact"—the only PT lumber sold at most lumber yards. In other words, stainless connectors are recommended unless you special-order lumber with lower chemical content.

8.5.7 Why is this still a secret after 10 years?

Many reputable contractors and engineers are not aware of the problem with copper-containing wood treatments. Most likely this is because they rarely get a chance to see connectors and fasteners after nature has taken its toll.

Members of the American Society of Home Inspectors have shared many of the photos included here that show extremely severe corrosion due to copper-based lumber treatments. ASHI members are on the forefront of seeing early failures of many different products. Lead paint and asbestos were used on millions of houses, but that does not make them any less toxic now that we finally acknowledged they are harmful. I expect that copper-treated lumber will eventually be viewed in a similar way—and I hope it's sooner rather than later.

Figure 8-53 *Left:* Anchor bolt from original 90-year-old construction.
Right: Anchor installed in the same crawlspace in PT lumber for an eight-year old remodel. The anchor in the PT material looks worse than the older one.
Photos: Justin Brodowski

Not even all *Hot-Dip* Galvanizing is Equal!

Most framing connectors are made from sheet steel that was galvanized before the steel was shipped to the hardware manufacturer.

Sheet metal parts are cut from huge rolls of HDG steel sheet. The cut edges of the parts are bare steel. Since the steel will not corrode if there is an anode in contact with it, the bare edge is protected—for a while. In the photos of corroded connectors, though, you will notice the red rust almost always begins to form near a cut edge of the material.

Many galvanized nails are made similarly—the nail manufacturer buys spools of HDG wire from a steel mill, then cuts the wire into short pieces to make the nails. The points and heads of the nails are not coated with zinc. Worse, the nail heads are formed by stamping the wire into a die so it flares out. In the sidebar explaining the chemistry of corrosion we saw that steel begins to corrode where it has been bent or deformed. Now we have a deformed area of the steel nail that also has no zinc coating on it—thus the nail head is usually the first place you see rust.

Other than stainless steel, the best readily-available finish to resist corrosion is hot-dip galvanized material. **The best HDG protection is provided by hot-dip galvanizing after fabrication.**

8.5.8 Solutions—or experiments?

Two methods to reduce corrosion without using all stainless steel connectors are fairly common, but not proven over the long term.

8.5.8.1 Mixing stainless steel and galvanized connectors does not solve the problem

For steel connectors and fasteners, copper is the enemy in PT lumber. But copper is not the only metal that will react with steel to cause fastener corrosion. You do not want to introduce another metal that begins to attack your connectors—yet combining stainless steel and galvanized fasteners has this result. Stainless steel is more noble than plain steel (see "The Chemistry of Corrosion" sidebar). If you use stainless steel together with plain or galvanized steel, the plain steel becomes the anode in the galvanic reaction. Now the plain steel corrodes because it is in contact with stainless steel; you could take the copper-treated wood completely away and the plain steel would still corrode.

Sometimes fasteners are mixed intentionally, sometimes accidentally. Installers may think that using stainless steel nails will improve the connection performance of galvanized framing connectors or foundation plates. The cost of mixing the two materials is certainly less than going all stainless steel.

Manufacturers clearly state that you should not mix different materials in the same connection.

8.5.8.2 Isolation membranes do not isolate the nails

For corrosion to occur, the two metals in question must either be in contact or at least have a way for small electrical charges to flow between them (water or moisture provides such a way). If the

---

Relative size of different metal parts affects corrosion rates

We know that whenever two different metals are joined, one of them will corrode—but how quickly? It depends on how much of which metal is present.

Say you have a two-pound steel foundation plate joined to *un*treated wood with stainless steel nails (adding another PT metal is more complication than we need). The foundation plate is the anode in this case, and will corrode until it is gone. Compared to the nails, the plate has a large surface area. If iron atoms in the steel plate are lost from the whole surface of the plate, it would take a long time for the plate to lose significant strength.

If we switch the metals so we have a stainless steel plate secured with steel nails, things get scary. Now we have relatively small steel nails acting as anodes that need to protect the much larger stainless steel plate; the nails will corrode very quickly. It could be that the same amount of steel is lost over the same time-span in this case or the one described in the preceding paragraph—but when the loss of a few small nails means a failed connection, the second installation is much more dangerous.

In the first case above, we cannot count on losing a thin layer from the whole surface of the steel plate. Some concentration of corrosion will likely occur right around the stainless steel nail.

two metals are kept physically separated, corrosion will not occur. So why don't we just wrap the evil PT lumber with a water-proof membrane to separate it from the metal parts? There are products that attempt to do this—but I do not see how they can actually work. This is because you cannot isolate the *nails* from the PT lumber. If only the nails corrode, you still lose connection capacity—but the nails provide a path for the electrical charges to flow, leading to possible corrosion of the "isolated" metal.

The membrane shown in Figure 8-54 is marketed specifically for decks, which are one of the main uses of PT lumber. At least you can usually access connections under decks to monitor the condition of the steel now and then.

8.5.8.3 Proprietary coatings

Major hardware manufacturers offer thicker zinc coatings on some of their products (for example, Simpson's "Z-Max" and USP's "Triple Zinc"). These coatings just provide a thicker layer of zinc (and since the parts are not galvanized after fabrication, all the cut edges are left uncoated). The thicker zinc buys some time—but the corrosion reaction is patient and unrelenting, so I do not trust a mere delay in connector death.

Figure 8-54 PT lumber separated from a sheet metal connector with a self-adhered membrane. The nails still penetrate the PT lumber. This method may delay corrosion in the connector, but the membrane does not offer any improved protection for the nails.

USP supplies *some* of their connectors with their "Gold Coat" finish. Gold Coat is "an organic top coat barrier layer" over galvanized steel. Coatings can get scratched during installation or handling; I simply would not want to trust my house to a thin coating separating essential connectors from corrosion doom. But I may change my mind after 50 years if actual field installations show good performance…

8.5.9 Is "long-lasting" long enough?

Many of the coatings and other protection methods and materials listed in the previous section were developed for outdoor decks. Compared to a house, the expected life of a deck is much less (as short as 10 or 15 years). What may be a satisfactory lifespan for a deck fastener would not meet most homeowners' expectations for the lifespan of their homes.

When a client with a 100-year-old house asks me to design an earthquake retrofit, I want to design a system that will last another 100 years. Copper-containing PT lumber poses a severe threat to that goal; I don't even want to rely on stainless steel, much less some unproven system—some of which have obvious flaws. Thankfully other chemical treatments are available that do not increase corrosion appreciably, described in the following section.

8.5.10 A better alternative: Borate-treated lumber

Another class of preservative treatment chemicals includes *borates*. The two common formulations are Disodium Octaborate Tetrahydrate ("DOT") and Sodium Octa-Borate ("SBx"). Borate treatments do not corrode fasteners even as much as CCA did, and therefore do not require special fasteners. Borates are not toxic to humans (this is the same class of chemicals as borax that you use to wash your clothes or hands and is a major ingredient in dishwasher detergent). Borates are much less harmful to the environment than copper and arsenic-containing wood treatments.

Unlike copper-based treatments, borates are colorless. Since carpenters are used to PT lumber having a green color to it, most borate-treated lumber is dyed light green or light blue. Figure 8-55 shows two samples of lumber treated with the same brand of borate.

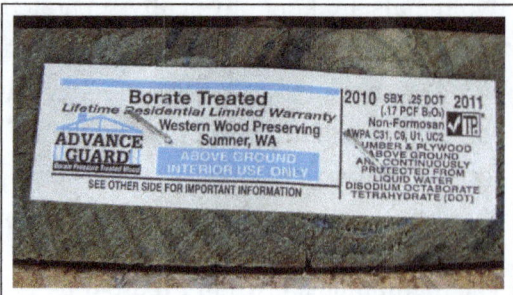

Figure 8-55 Borate-treated lumber does not corrode fasteners; it is also not toxic to mammals. Borate-treated lumber cannot be exposed to liquid water, but is appropriate for most retrofit applications. It is also usually less expensive than wood treated with toxic chemicals. California Cascade Industries (phone 800-339-6480) supplies borate-treated lumber throughout California. A few lumber yards in the SF Bay Area [Osborne Lumber (in Newark), Sierra Point Lumber (Brisbane) and Golden State Lumber (several locations)] carry borate PT lumber as a regular stock item in limited sizes and lengths. Special-ordering from other yards is very worthwhile—but see the warning in the sidebar, next page.

> **Borates do not kill termites directly**
>
> Termites do not actually digest wood—they only chew it up and swallow it. A variety of fungus that inhabits termite stomachs breaks down the wood. The termites get their energy from the fungus. Borates kill the fungus—thus the termites die of starvation.

8.5.10.1 Limitations for borates

Borates will dissolve in water, so borate-treated wood must be protected from liquid water—this rules it out for building decks or fences. Borate-treated wood is perfectly acceptable for use in foundation sills.

Borate-treated lumber is also not allowed in areas where Formosan termites are present (Hawaii and Florida, possibly other Gulf-coast areas).

8.5.10.2 Why don't more lumber yards stock Borate-treated lumber?

As with the two treatment levels used for copper-containing PT lumber, suppliers do not have space for yet another type of material. In most suppliers' minds, there is no advantage to borate treatments—borates are not water-proof, and the suppliers assume the products they already stock will work in all situations. Some people in the construction industry are also under the false impression that borate-treated wood cannot be used for mudsills because borates are not waterproof.

8.5.10.3 Other benefits of borate-treated lumber

One advantage of borate-based treatments is that they will travel toward moisture in the wood, which is exactly where you need them. In log cabin construction, large borate crystals are placed in holes bored in the logs. If moisture reaches the crystal it begins to dissolve. The borate moves through the moist wood, spreading out away from the crystal. In this case the fact that borates will move through the wood is helpful.

According to the borate-treating companies, incising is not required for the treatment to adequately penetrate the wood. If true, this means you do not need to spot-treat cut ends of boards.

For more information on borate treatments, search for:
"Timbor" by US Borax
"Sillbor" by Arch Chemicals
"Advance Guard" by Osmose

Warning: *check the lumber yourself!*

In 2005 I hired a contractor to build an addition to my house. I told him to make sure he put in a special order well in advance for borate-treated lumber to use for the mudsills. The contractor and his crew arrived to lay out walls the day before the lumber yard was to deliver all the framing lumber—he assured me that the yard had ordered borate-treated stock for the mudsills.

The lumber delivery arrived the next morning, including a dozen or so dark green boards for mudsills. I checked the lumber end tags: ACQ. Those boards went back to the yard. I took the morning off, drove an hour to the nearest yard that had borate-treated lumber in stock, and returned with what was needed so the crew could begin building a structure that would not self-destruct.

Onward to 2014; I designed a foundation replacement and earthquake retrofit for a 1915 bungalow. I referred the owner to a premier foundation contractor. The references on the drawings to borate-treated lumber got their attention, and they special-ordered borate-treated mudsill stock. When I went to crawl under the house to observe the work before concrete was poured I immediately noticed dark green mudsills; no end-tags were visible, so I went to the lumber yard the next day. The only PT lumber they carried was ACQ, and they told me their supplier filled the "special order" with ACQ.

Lesson: If you've read this far, you know more about PT lumber than most lumber suppliers, engineers, contractors, and building inspectors. Don't rely on them.

8.5.11 Another alternative: Isolate untreated wood with water-proof membrane

The building codes generally call for foundation sills and posts resting on concrete to be of PT lumber or decay-resistant wood (IBC/CBC Section 2304.11). In some specific cases, untreated lumber is allowed if an "impervious moisture barrier" is used between the wood and concrete (such as posts in metal post-base connectors). The spirit of the code seems to allow use of untreated wood if it is separated from concrete by a moisture barrier. If separating untreated mudsills or other members from concrete was more common, the code would probably include a specific provision allowing this practice. The absence of such a provision does not mean a membrane would not protect the wood.

Figure 8-56 shows the beginning of a wall frame isolated from a concrete stemwall by self-adhered flashing. The flashing should block moisture from the concrete, thus protecting the wood from decay. Note that "tar-paper" does not block water vapor effectively and should not be used.

Figure 8-56   Untreated lumber shown here is protected from moisture wicking through the concrete foundation wall by "peel-and-stick" membrane (also called self-adhered flashing, commonly used to waterproof around window and door openings). Note poor bolt installation.

I do not like to give termites a path from the ground to untreated wood without crossing borate-treated lumber somewhere. However, I would take my chances with the system shown rather than use copper-containing treatments that I know will corrode fasteners

8.5.12 Recommended best practice for known damp areas: All of the above, short of stainless steel

For the amount of trouble it is to install most earthquake retrofits, you may as well spend more on good materials to provide a superior system. Use borate-treated lumber, hot-dip galvanized connectors (if nothing else, at least the nails should be hot-dipped after fabrication), and an isolation membrane. The membrane will keep moisture from wicking into the wood structure. Even foundation grade redwood decays after decades of contact with damp concrete; the borate-treated material may not decay, but it could transmit moisture to other wood components.

Chapter 9 Special Structural Concerns

9.1 Hillside homes

"Hillside homes pose, without question, the greatest life safety risk of any type of damaged dwelling investigated by this subcommittee." From "Findings & Recommendations of the Residential Cripple Wall Subcommittee of the City of Los Angeles and SEAOSC," from Ref. #11, Appendix I

Figure 9-1 Houses on steep hillsides, especially houses on unbraced "stilts," perform very badly in earthquakes. FEMA photo

Hillside homes offer spectacular views of the surrounding areas. They often perform worse in earthquakes than homes on flat or gently sloping sites. The main reason is the uphill edge of the house is directly attached to a foundation, but the downhill edge has relatively little bracing. This leads to the main floor of the house twisting back and forth, perhaps tearing completely away from the foundation, causing almost total destruction as shown in Figure 9-1. Houses on "stilts" can be even more vulnerable than houses where the space below the main floor is enclosed.

Figure 9-2 Steeply sloping or stepped shear walls cannot function the way we expect them to. Special measures are needed to brace either of the walls shown. *Left*: Pier and grade-beam foundation (piers are underground, not visible; concrete grade beams meet near the middle of the photo). *Right*: Stepped foundation with "rat-slab" on right side of photo

Besides the twisting action mentioned above, stepped or sloped shear walls constructed on hillside foundations will not perform as expected. Figure 9-2 shows two typical hillside foundations with wishfully-installed plywood panels. Shear walls must be rectangular to function the way engineers expect them to. (For further information refer to the Shear Wall Guide—see Appendix I).

9.1.1 When does a hillside home become especially hazardous?

As usual, many factors affect how badly a house might perform during an earthquake. Consider the following:

1. How steep is the site? If the ground slopes more than one story from one side of the building to another, plywood bracing walls rapidly lose their effectiveness. The house in

Figure 9-3 slopes so much that the "crawlspace" is two stories tall on the left side of the photo.

2. Is the bottom of the house enclosed, or is it on stilts? The house shown in Figure 9-4 is in danger from side-to-side shaking. Diagonal bracing can provide stability, but long, spaghetti-like braces like those shown in the photo are generally not adequate; do not assume that some spindly boards nailed into place will protect you. The house shown in Figure 9-5 has no bracing at all. The extension of the house beyond the downhill supports will increase the proportion of earthquake forces that will act on that row of supports.

3. How is the house configured? Look again at Figure 9-3. What may look like a penthouse is a two-car garage constructed as the top level of this tower. The additional 3-inch thick concrete floor slab (typical for garage floors over the last several decades) will double the weight of the top level of this house—the only worse place for the added weight would be the roof.

4. Does the house have other vulnerable conditions? Combinations of split-level construction, weak wall lines, and so forth mean even greater risk.

5. How sturdy is the foundation system? Isolated concrete pads can move relative to one another; continuous perimeter foundations with "conventional" stepped foundations are better; "pier and grade beam" foundations are the best, and cost accordingly.

Figure 9-3 Typical 1960's era house in the East Bay hills; garage with heavy concrete floor is at the very top level.	Figure 9-4 Portion of house on lightly-braced stilts.	Figure 9-5 House on unbraced columns; extension of floor past the supports will magnify the earthquake forces on them.

9.1.2 Solutions to hillside conditions

In some respects, hillside construction offers good opportunities for making very secure connections from the floor framing directly to the foundation system, at least on two or three sides of the building.

If you have a hillside home I strongly recommend that you consult an experienced engineer to determine appropriate strengthening measures. The following sections show methods that were appropriate for *specific* houses; these methods may apply to similar houses, but not necessarily all situations.

9.1.2.1 "Hill-Side Anchors" at sloped or stepped side walls

Figure 9-6 shows an all-thread rod connecting floor framing members to the highest portion of a concrete foundation. "Hold-down" brackets (installed horizontally in this case) connect to the cripple wall top plate or concrete at either end of the rod. (Note: Earlier varieties of hold-down brackets offered some minor adjustment in placing fasteners into wood members. Current hold-downs may not allow installation of screws with sufficient edge distance to cripple wall top plates; I recommend installing an LVL or similar member to the face of the cripple wall before attaching the hold-downs, which allows better placement of fasteners.)

Figure 9-6 *Above:* "Hillside primary anchor" from foundation at right to top plate of cripple wall at left. Shear transfer ties will be installed to connect joist blocking to the top plate. Plywood bracing is not needed on the cripple wall, nor are mudsill anchors to the stepped foundation.
Right: Close-up of worker installing tie-rod into brackets for connection to foundation.

In the connection shown in Figure 9-6 it might have been possible to embed the all-thread into the end of the top step in the foundation, setting the rod with epoxy. I do not trust chemicals like epoxy to last for several decades or longer before an earthquake demands their full strength. In this case, face-mounting the hold-down brackets to the inside of the foundation allows connecting them with bolts that will act in shear; this strikes me as a much more reliable connection.

One advantage of the hillside primary anchor is you do not have to worry about cripple walls below it. The anchor delivers all of the earthquake force from the structure above directly into the foundation, so there is no need to add plywood bracing panels to the cripple wall.

Some houses already have most of the elements of hillside anchors that are just waiting to be connected. The green beam at the top of the right-hand photo in Figure 9-2 is an enormously strong member that, if connected properly, would serve well as a primary anchor.

Hillside primary anchors are required in new construction in Los Angeles. The primary anchors attach to both ends of floor diaphragms. "Secondary" anchors are also required at intervals along the uphill edge of the floor diaphragm (see Figure 9-7). Houses that cascade down a hill need both primary and secondary anchors at each floor level.

Many custom houses built in the last 35 years have "pier and grade-beam" foundations that are very solid. Unfortunately most of these houses have stepped or sloping "not-really-shear" walls and would benefit from hillside anchors.

Figure 9-7   Hillside secondary anchor; note that this serves a very different function from the much-maligned "angle-iron" anchors in Section 8.2.1.

9.1.2.2 "Floating" shear walls

If you build a shear wall on a sloped or stepped foundation, the bottom of the shear wall is not level and thus you do not really have a shear wall. Why not construct an elevated base for the shear wall? If we provide a continuous member along the bottom of the plywood panels, they will not know the difference between a mudsill or an elevated strut. Figure 9-8 shows such a wall.

A tie-down bracket oriented horizontally provides the "reaction" force at the base of the wall. If the plywood panel was taller, tie-downs at both ends of the shear wall would be needed to resist overturning forces. Such a case is shown in the detail in Figure 7-43.

9.1.2.3 Custom steel brackets

Some existing configurations are difficult to adapt to "off-the-shelf" hardware. Welding structural steel members together into custom assemblies may be the most economical choice in some cases. Figure 9-9 shows a carefully made bracket. Designing brackets like this is best left to an engineer, as the forces

Figure 9-8 Floating shear wall; bottom edge of plywood is connected to the horizontal 4x8 which is in turn anchored to the grouted masonry foundation.

Figure 9-9 This custom-made steel bracket secures the floor framing system directly to the foundation (note that the small, gray steel connector near the center of the photo is an obsolete "hold-down" that may or may not be installed in a useful manner).
Photo: Paul Barraza, GGASHI

and interactions between different components, and connections to the framing and foundation, can get complicated quickly.

Current code requirements are very restrictive for welded steel. Specific quality control and inspection procedures must be followed. Steel is 10 to 20 times stronger than wood, so while the inspection requirements make sense for hospital construction they seem inappropriate for wood-framed residences—especially considering that inspection costs could easily double the steel fabrication expense.

9.1.2.4 Walls perpendicular to the slope

As shown in Figure 9-10, walls on the downhill side of hillside homes can exceed 16 feet tall. A 16-foot tall shear wall could deflect several inches under earthquake loading. The floor framing may connect directly to the foundation at the uphill side. If one side of the main floor shifts a few inches and the other side is fixed directly to a rigid footing, the floor framing rotates like a giant lazy Susan. The rotation may be small, but it does not take much movement to rip connections apart. As the earthquake shakes back and forth the floor twists and pulls loose from the foundation. There are two common ways to reinforce against rotation.

Stiffen the downhill wall so it will deflect about the same as the uphill wall under the same load. (If the main floor attaches directly to the foundation at the uphill side of the house, this is not an option.) Shear wall deflection depends on the following:

Figure 9-10   Wall on left at downhill side of house is 16 feet tall; opposite side of the floor diaphragm connects directly to the foundation, which would result in the floor twisting during a quake and possibly tearing loose from the footing.

- Length of the shear wall
- Height of the wall; If you double the height of the wall, the deflection also doubles
- Nail size and spacing; deflection is less for bigger nails and closer edge-nail spacing
- Type and thickness of plywood
- Amount of slip in tie-downs, if they are used

To make the downhill wall sufficiently stiff to match the uphill wall, you may need to use any combination of closer edge-nail spacing, thicker plywood, bigger nail diameter, and longer wall length than you would need for strength alone.

Ignore the downhill wall altogether and provide for the entire base shear connection along the uphill edge of the main floor framing to the foundation. This method uses "diaphragm rotation" that is sometimes applied to open-front buildings (see Section 9.2.1). You must make sure that the rotation can be resisted by the connections in the direction the hill slopes (this is a problem if you brace only the short side of a long, skinny rectangular diaphragm).

9.2 Soft- or weak-story construction

The terms "soft story" or "weak story" are often used to describe construction above a garage door or large expanse of windows. Soft or weak story construction performs poorly in earthquakes, as show in Figure 9-11.

A more collective term is "soft, weak, or open-front" (SWOF) construction.

SWOF construction is very common in San Francisco; in other cities, single-family homes are less likely than multi-family buildings to have SWOF problems. Estimates put the number of SWOF houses in San Francisco at 25,000.

Figure 9-11 An earthquake tore this house in half. The segment on the left used to have a garage under it, but the garage walls collapsed. The wall with sunlight on it shows the outline of the roof where the two segments originally connected. The left portion of the house fell about seven feet. FEMA photo

Some SWOF buildings, if built new, could be built with prefabricated shear walls such as Shear Max, Strong Walls, Z-Walls, Hardy Panels, etc. Installing such products in a retrofit is usually not an option without a serious foundation upgrade, though, because of the huge uplift forces involved. Old footings usually come pre-cracked and unreinforced; installing a pre-fab shear wall just gives the earthquake a big lever to rock the footing out of the ground.

There are two general ways to strengthen SWOF construction. One is less costly but offers less protection than the other. The "easy" method requires bracing on only three sides of the floor diaphragm above the soft story. The current building code allows this method for new construction (with serious restrictions; see Section 4.2.5.1.1 in SDPWS (Reference). FEMA recommends against using diaphragm rotation to resist earthquake forces.

The better, and almost always more expensive method is to provide some sort of earthquake resisting element along the edge of the diaphragm above the SWOF condition.

Soft Story Technicalities Engineers use "soft story" and "weak story" to describe very specific structural conditions. "Soft-story" refers to construction where one particular story has less than 70 percent of the *stiffness* of the story above. "Weak-story" is the case where any story has less than 80 percent of the *strength* of the story above. For wood-framed construction, strength and stiffness are closely related and the distinction between weak-story and soft-story is rarely important. Technically cripple-walls are usually a soft or weak story, or both. Open-front buildings may not meet the code definitions for soft or weak stories.

Soft Story Triage Garage doors and large windows attract attention, but may not always indicate serious concern. The following show general cases of extreme risk to low concern.

Figure 9-12   Severe soft story, heavy stucco, room extends significantly from house. "L" shaped build-ings tend to break at inside corners.
Photo: Roger Robinson, GGASH1

Figure 9-13   Slightly less bad than house at left: weak wall does not extend as far from main house. Split-level floor separates the garage wing from portion of the house to the right of the entry, preventing structural connection between wings.

Figure 9-14   Still worth strengthening. Garage door flanked by walls a few feet wide will fare better than houses shown above. Wall of windows above garage door creates a weakness at main floor, but at least there is not the weight of another floor above that level.

Figure 9-15   Not a major hazard. Large wall with no jogs or changes in floor level means that this building has lots of reserve strength around the garage opening.

9.2.1 Diaphragm Rotation: the budget solution

Up to this point we have assumed that all four sides of a square or rectangular floor system need bracing for the system to resist earthquake forces. You can actually get by with braced walls on only three sides of the building, under certain conditions. The diaphragm rotation transfers all of the lateral force to the rear wall of the garage. The earthquake force acts back and forth through the center of the floor platform; resistance is provided at the rear wall, and the two off-set forces tend to cause the floor system to rotate. The side walls of the garage resist the rotation. For new construction the building code limits using this method to certain ratios of depth-to-width of the diaphragm.

> Diaphragm rotation is likely the unintended structural principle that keeps many garages from collapsing, even though they have narrow shear panels on either side of the garage door. Narrow shear walls deflect a great deal before they begin to resist appreciable load. To make matters worse, tie-downs often loosen up because of wood shrinkage. In many cases the rear wall of the garage is resisting the great majority of earthquake force delivered through diaphragm rotation.

9.2.2 Install a structural system at (or near) the front of the garage

Ideally a floor or roof diaphragm will have some sort of earthquake-resisting element along each edge. Sometimes moving the attachment slightly away from the edge is needed because of various possible restrictions.

9.2.3 Moment-resisting steel frames

Also called "rigid frames" or simply "*moment-frames*," moment-resisting frames are thought by some to be the Cadillac solution for SWOF conditions. The strength of a moment-frame is at the connection between the columns (vertical members) and the beam. Figure 9-16 shows a moment-frame and a close-up of the beam-to-column connection. The configuration of the connection that looks like an inset square is the tell-tale sign of a moment-frame. Without this special corner connection (or something similar in the case of manufactured frames that are delivered unassembled and bolted together on site) the frame has very limited ability to resist side-to-side forces. Figure 9-17 shows two frame connections that are intended to support only vertical loads; square or round columns are often used to support heavy loads,

Figure 9-16 Moment-resisting frame, or simply "moment-frame" Left: Frame bracing front wall of building above parking area. Right: Close up of corner joint.

195

Figure 9-17 Examples of steel connections that will NOT resist appreciable earthquake forces.
Left: Gaps between the column and the top and bottom flange of the beam cannot transfer bending forces to the column.
Right: Pipe columns have very little strength to resist side-to-side earthquake forces.

Figure 9-18 Moment-frame joined with bolts at the middle of the beam. While shipping the frame in two pieces was easier, tall and heavy upside-down "L" shaped items are hard to stand in place.

but they are poor at resisting lateral loads. Just because you have steel columns, or even a complete steel frame, does not mean that it was designed to resist earthquakes. The vertical and horizontal members in a moment frame are typically all made of "wide-flange" steel stock.

A couple of manufacturers like Simpson and Mitek make a few "off-the-shelf" moment frames, and will also build them to custom dimensions. Retrofit installations almost always require custom-sized frames—if a frame will work at all. Steel fabrication shops can also assemble frames and usually offer the option of delivering and installing them in the field. Figure 9-18 shows a frame that was fabricated in two "L" shaped sections for easier transportation to the jobsite; the sections were bolted together in the field and the existing wood framing was secured to the steel members with structural screws. The existing building almost always prevents using a crane to install new members items in existing buildings. The "L"-shaped assemblies were awkward to stand into place from the ground.

Problems with moment frames: Moment-frames are expensive. The frame shown in Figure 9-18 cost $11,000 picked up at the fabricator's shop (in 2010, when fabricators were hungry). Simpson "Strong-frames" start at about $6,000. Engineering for a moment-frame adds another layer of complication to a retrofit. Moment-frames must be designed by an engineer.

Besides the relatively high cost of moment-frames, they often just won't fit into a retrofit. Usually the SWOF portion of a building is at the front; this is also where the electric, gas, and water service come into the building and sprawl on the inside of the front garage walls like ivy. In San Francisco, where there is usually no side yard, there may also be a stairway or access door in the front wall in addition to the garage door. Often the garage door is so close to at least one side wall that the vertical frame members will not fit without make the garage door narrower. In many cases the garage ceiling height is only seven feet or so, and reducing the headroom by a foot is out of the question. Luckily engineers have another SWOF solution in their bag of tricks: the moment column, described next.

> **Steel fabrication notes for engineers**
>
> Welding is *really* expensive. Small, voluntary retrofit projects cannot absorb welding and inspection costs the way a hospital, school, or high-rise project might. (For a recent moment-frame installation, steel fabrication and installation costs were about $6,000. Inspection and testing costs added $3,800.) Whenever possible I use moment-columns and connections that do not require any welding.
>
> Connecting steel to wood is often done with "Nelson studs." These require welding, and then careful location of drilled holes in the wood member. A much better solution (less expensive and more reliable connection) is just to provide holes in the steel for structural screws to connect through to the wood. Welded studs can also make assembly more difficult, as when installation requires slipping a new steel component in between existing wood framing members..

9.2.4 Moment-columns

Moment-columns, also called *cantilevered columns*, are often the only viable solution for an existing SWOF structure. Think of a moment-column as an extremely rigid flag-pole anchored to a massive foundation. The top of the column connects to the floor framing above the garage and provides stability in an earthquake.

Figure 9-19 shows a moment-column set between two garage doors. Since the garage doors extend to within inches of the side walls of the garage, installing a moment-frame would have required making at least one of the doors narrower. Most owners do not find this an acceptable choice.

Figure 9-20 shows a sequence of a moment-column installed to fit between obstructions.

Figure 9-19 Moment-column between two garage doors. The floor slab was cut out to allow digging a new foundation trench to hold the column base.

Figure 9-20
Left: Garage floor slab cut out, footing trench dug with concrete reinforcing bars in place. Base of column will be embedded in concrete.
Right: Top of column connected to floor framing of room above using a simple system of bolted-on structural steel angles and accessories.

9.2.5 Wood moment-resisting frames (aka portal frames)

The top diagram in Figure 9-21 shows a generic foundation with two narrow shear walls on it—like many garages have. (Engineers will note that this is not a complete "free-body diagram.") Mudsill anchors transfer shear to the foundation; connections at each end of the panels prevents them from rotating like square wheels. For our purposes, the structure would work the same if we could turn it upside down. The system shown in the bottom diagram in the figure is structurally identical to the shear wall-on-footing system—the "hold-down" forces are now applied at the tops of the shear walls, provided by connections to the beam.

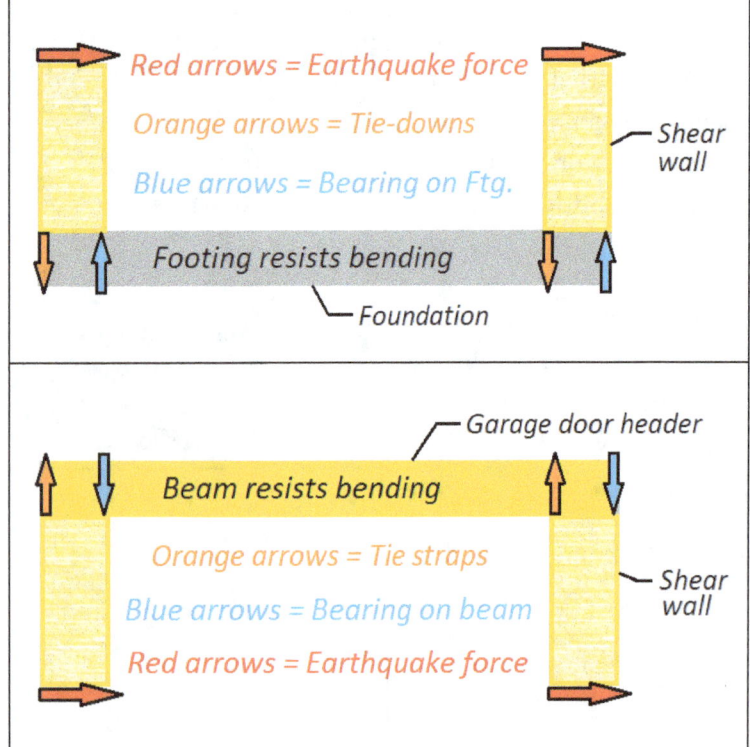

Figure 9-21 Diagrams comparing shear walls to wood "moment frame."
Top: Shear wall segments resist earthquake forces (red arrows) applied at the tops of the shear wall segments. Overturning force is resisted by tie-downs (orange arrows). End of shear wall opposite tie-down bears on foundation (blue arrows)
Bottom: Garage door header spanning full width of front garage wall secures the *tops* of the shear wall segments. In this case straps from the shear wall end posts to the beam (orange arrows) together with bearing against the beam (blue arrows) prevent the shear wall segments from overturning.

Some garages were constructed with a continuous header that runs the full width of the front of the building—Figure 9-22 shows an interior view of one such case. If you are lucky enough to have this configuration in your garage, you are most of the way to the "upside-down" shear wall system shown at the bottom of Figure 9-21. The header serves as the member to which you connect "tie-downs." In this case the tie-downs are simply "ties," but they still act to prevent the narrow shear wall from rotating. Not much else changes—you still need to nail each edge of every shear panel; the mudsill connection and roof sheathing connections transfer forces to and from the shear wall, and so forth. But in this case you do not have concentrated uplift forces trying to lift your footing out of the ground. (The entire wall has the same overturning forces applied to it, but is so much wider that it takes much less force to resist the overturning.)

Figure 9-22 Inside view of narrow shear wall next to a garage door. The header spans the full width of the garage, allowing straps to connect from posts on both ends of the shear wall. The narrow plywood panel on the outside of the framing resists horizontal forces.
This arrangement creates a "moment frame" out of wood.

Wood moment-frame proves itself

Just before I graduated from college, I bought a home in a planned unit development in Paso Robles, CA. The house has a concrete tile roof. Tile roofs add a lot of weight to a house, and increase the earthquake forces on them. About 60 other units in my development were virtually identical; the garages have narrow walls on either side of the garage door. (Figure 9-22 shows the inside of the garage at the top of one of the narrow walls.)

I installed the straps shown in the figure, repeating the process on the other side of the door; the hardware cost less than $20. The most time-consuming part of the installation was tightly shimming the gap between the header and the single top plate below it where the straps cross from studs to beam.

Eventually I moved three hours away and rented the house to a professional roofer. On December 22, 2003, the San Simeon earthquake hit. I called my tenant just after the earthquake to see how things were. He marveled, "It's really weird—the roofing tiles along the ridge of every other garage on the block were thrown about three feet to one side, but not a single tile moved on this house!"

9.3 Irregular and unusually shaped buildings

The strongest shape found in nature is a sphere. The strongest practical shape for a house is a cube. Putting a sloping roof on a box still gives you a strong building. Problems arise when you join boxes of different sizes and shapes, or if you cut out portions of the box. Houses that stray too far from uniform box shapes perform badly in earthquakes. Building codes address several "irregularities" that involve a building's footprint shape, profile shape, stiffness, weight distribution, and other factors.

9.3.1 Plan irregularities

Plan irregularities refer to the shape of the building's outline when viewed from above. If the overall floor outline of a house looks like an L, T, U, H, or some other configuration besides a square or rectangle, different wings of the house will tend to move in different directions during an earthquake. For wood-framed construction the worst that results from this is usually localized damage, rather than partial or complete collapse. Large openings in floor areas, such as two-story foyers that cut into the main second-floor diaphragm, can create plan irregularities.

9.3.2 Vertical irregularities

Buildings that change height from one portion to another are said to have a vertical irregularity. Houses with single story wings adjacent to a two-story wing are a common example of this irregularity. Essentially this means houses that look like an L, U, upside-down T, etc. when viewed from the side. Vertical irregularities are also unlikely to lead to severe structural damage (collapse of the building or a segment of building) in typical wood-framed construction.

9.3.3 Weight irregularities

For residential construction, weight irregularities are fairly rare—the code definition relates to significant changes in weight from one floor level to another. Very heavy roofs or roof decks can create a weight irregularity if the floor level below is much lighter construction. In some hillside houses the garage is at the top level of the house, where a concrete garage slab roughly doubles the garage's weight compared to similar construction with a wood-framed floor. In areas where heavy snowfall is typical, snow on roofs or decks adds an amazing amount of weight beyond the structure itself. Building codes typically require increasing the design forces to include the effects of snow on top of your house shaking around during an earthquake.

9.3.4 Strength and stiffness irregularities

Cripple walls typically create both soft story (stiffness irregularity) and weak story (strength irregularity) conditions. Soft, weak, or open-front ("SWOF") construction is a major concern but cannot be addressed in this book.

9.3.5 Combinations of problems

Usually any one of the irregularities listed above would not pose a safety hazard. When combined, however, they can quickly pose significant hazards. Houses on down-slope hillside lots often include several irregularities.

Chapter 10 Foundation Problems; Cracks, Termites, Etc.

Earthquake forces travel throughout the house, increasing toward the foundation. Old houses can settle, deteriorate, and get eaten by pests. This chapter addresses some common structural concerns that relate to earthquake resistance; in my opinion, some of these issues get more attention than they deserve.

10.1 Old concrete

Old concrete does not always equal bad concrete, just as new concrete does not always equal good concrete. Another experienced structural engineer advises that if the concrete does not fall apart as you drill into it to set foundation anchors, it is strong enough to provide adequate earthquake resistance. (Note that this assumes you are connecting *wood* elements to the foundation, which by comparison are considerably weaker than steel or concrete). Whether the concrete will be strong enough in 10, 20, or 50 years from now when a considerable earthquake strikes is another question.

Replacing old foundations adds enormous cost to retrofit work. Foundations should be replaced if they are bad—they should not be replaced if they are just "old." This was addressed in Section 4.1.1.

10.2 Settlement

Almost all houses settle. If you want a house that has not settled, buy a new house. After 50 or 80 years, it will have settled—if only a little.

10.2.1 Effects of settlement

For wood-framed houses, *generally speaking*, settlement hardly ever causes a structural hazard—though it may be a serious cosmetic concern. The barn in Figure 10-1 shows how much settlement a wood-framed structure can endure without collapse. Settlement by itself rarely poses a threat to the structure and rarely interferes with earthquake retrofit work. However, retrofitting a house that has uneven floors will lock things into position so it becomes very difficult to restore the structure to its original condition without removing the retrofit system.

Figure 10-1 Wood-framed structures can tolerate an impressive amount of deformation without collapsing.

Wood members will "creep" over time to adapt to slowly settling supports. The members gradually relax into the settled position and "set" in their deformed shape. If you quickly re-level a wood structure that took decades to adopt its current shape, it will take a year or more for it to adjust to its restored configuration.

Ideally you could put a whole new steel-reinforced concrete foundation under your house, but if you cannot afford to replace your foundation don't fear that settlement puts you in certain danger.

> I worked in an office in a converted 1894 house for a dozen years. The room where my boss had his office had settled so much that it sloped a full 7 inches across its width; when he got tired of the wind blowing through the gap that opened between the floor and the baseboards, he had the building re-leveled and a new foundation installed.

10.2.2 What caused the settlement?

If the settlement has occurred over several decades, usually I tell clients not to store their marble collections on the floor, and enjoy the quirks of an old house.

Usually settlement is caused by soil "self-compacting" under loads as small particles gradually reposition themselves when moisture travels through the pores in the soil. Settlement due to self-compaction usually slows over time as the voids in the soil fill in or compress. Resulting movement is very gradual and diminishes with time.

Gopher and rodent activity, or even ants and earthworms, will remove soil from under your foundation. Soils may shrink and swell seasonally with changes in moisture levels. Roots will grow under your house, lifting it slightly, then die and gradually leave cavities in the soil as they decay. Avoiding these natural actions requires deep foundation systems that extend down to soils below roots and critters—or in some cases, below uncompacted soil fill.

10.2.3 When to worry

As with any structural concern, there is no universal rule that applies. Brittle wall finishes such as stucco or brick do not tolerate settlement. Even if the wood structure is not in danger of failing, cracked stucco can allow significant water intrusion and resulting decay.

The following conditions warrant further investigation:

- **Sudden loss of support for other structural members:** Figure 10-2 shows a post that has settled significantly and no longer supports the roof beam. The settlement occurred fast enough that the beam did not "creep" downward and maintain contact with the post; this could indicate a soil problem that might rapidly lead to serious damage.

Figure 10-2 Support for this post was lost quickly, allowing the gap to form.

- **Settlement caused by soil slippage:** If part of your lot is sliding out from under your home, that is a huge concern compared to gradual, declining movement. You will need to consult a soils engineer or engineering geologist.

- **Landslide movement:** The house shown in Figure 10-3 was built on a slow-moving land-slide (a "mud glacier" is the description given by soils engineers). Strengthening the foundation of this house will be a major undertaking. Landslide movement can result in door and window openings in walls that run in the direction of the slope gradually deforming. Other clues include leaning utility poles or retaining walls (though other soil conditions can cause these) and cracks in pavement. *Peace of Mind in Earthquake Country* by Peter Yanev illustrates some of these conditions. One of Berkeley's premier soils engineers has generously provided a map of landslides in the Berkeley hills here:

  Figure 10-3 Landslide movement will overpower most standard foundations and structures. A slow-moving landslide is gradually tearing this house apart.

  http://www.akropp.com/wp-content/uploads/2013/07/berkeley_hills_slide_map.pdf

  Perhaps residents in other areas are lucky enough to have similar resources.

- **Settlement caused by erosion:** Soil worn away by running water (including water from plumbing leaks) will lead to continual and increasing loss of structural support. On steep hillsides, gradual erosion caused by deer, horses, dogs, or other animals is common along the downhill side of hillside homes.

- **Brittle finishes susceptible to damage:** If your house features tile floors, brick, stone, etc., such materials are much less tolerant of loss of support than wood flooring, carpet, or drywall. This is more of a cosmetic concern than a structural one—but most people will rightfully get upset about $20,000 worth of cracked Italian marble.

- **Movement due to frost heave:** Not much of a concern along the west coast of the continental US, frost heave is a continuing problem that will eventually succeed in damaging something serious. If you are in an area with frost heave you should consult a local expert—not one from California…

- **Footing rotation:** Settlement can lead to footing rotation (addressed in Section 10.4) which may need to be addressed before you can install a sturdy retrofit.

10.2.4 Settlement's effect on an earthquake retrofit

From an engineering standpoint, if a foundation can remain stable and hold together during an earthquake, it does not matter how ugly it is. When shallow footings led to settlement that may be ongoing, I either recommend stabilizing the footings with "bench piers" (in rare cases that this is the preferred solution) or replacing them altogether. Full replacement is often not much more work than installing bench piers and makes a much better impression on future buyers of the house.

Keep in mind, your last chance to make significant foundation repairs is before or during the retrofit process. Typical retrofit work will pretty much lock the structure into its current position. Without cutting everything loose from the framing above, it will be impossible for anyone in the future to re-level retrofitted walls.

10.3 Cracks in concrete foundations

Concrete gains strength by "curing," which is the chemical reaction between water and cement. Concrete is generally considered fully cured after 28 days, meaning it does not reach its full strength for about a month. Curing does not occur because the concrete dries out; in fact, curing is best accomplished by keeping the concrete moist. As concrete dries it shrinks. If it dries out and shrinks faster than it gains strength from curing, then cracks form.

"Small" cracks are generally inconsequential. Small is a relative term. For concrete placed before 1940 or so, a crack has to be wider than my index finger (about ½") before I get very excited about it. Cracks in concrete that was placed after 1940 start to draw my attention when they exceed ¼" width (about the size of a pencil).

Figure 10-4 Typical shrinkage crack; this crack likely began forming within hours of when the concrete was placed, and grew over a year or two until the concrete dried out to reach equilibrium. (The mudsills were enough to strengthen the concrete and force the infant crack to form at the mudsill joint—a common occurrence in level foundations.) Shrinkage cracks are rarely a cause for concern.

10.3.1 Cracks to expect

Shrinkage cracks are almost impossible to prevent. You can control *where* they occur, but they **will** occur unless you take extreme precautionary measures. In concrete or masonry walls with uniform height, vertical cracks such as the one shown in Figure 10-4 generally result from shrinkage. In stepped foundations you almost always see diagonal cracks at each place the top of the stem wall steps up to the next highest level. The cracks are lazy, and follow the shortest length from the top of footing to the bottom; in the case of a stepped footing the shortest distance is where the step has already created a notch in the wall.

Cracks also appear at the inside corners of openings for doors, windows, crawlspace access, etc.—any place that an otherwise uniform wall or slab has been weakened. The cracks are usually diagonal. They will always follow the weakest route, though, which may not be a diagonal path.

Steel reinforcing, or other strengthening (or weakening) factors can redirect cracks in directions contrary to the above generalities. I have seen footings that were "reinforced" by the redwood mudsills: each joint in the mudsill had a corresponding shrinkage crack in the footing under it—the cracks were small, but they did not occur anywhere else.

10.3.2 Cracks to worry about

As many home inspectors succinctly state in their reports, "most large cracks were once small." How might you know if a small crack is likely to get larger, or if a large crack is a sign of ongoing distress? The best way is to arrange a site evaluation visit from an expert in such matters who will consider soil properties, loading conditions, concrete quality, the possible presence of reinforcing steel, and other factors. They will likely be looking for the following, and more:

- Is the face of the concrete offset from one side of the crack to the other? This shows that the concrete on one side of the crack is moving relative to the other. If the face of the concrete remains smooth across the crack there is usually less concern that the foundation is shifting.
- Is the crack wider at one end than the other? This usually means the foundation is sinking at one end or the other of the separate segments. Note that hairline shrinkage cracks often trail off until they disappear, but they usually start off being relatively narrow.
- Water flowing through cracks is a bad sign. Although not a likely result of foundation problems, moving water could certainly lead to them.
- Horizontal cracks may indicate basement walls that are bulging inward, or that a "cold joint" in the concrete is present (a cold joint occurs when the first layer of concrete is allowed to harden before the next layer is placed—if there is no reinforcement across the joint, the concrete tends to break at that location).
- Cracks at the corners of older foundations often indicate that the footings are rotating (see next section)

10.4 Rotated footings

In the SF Bay Area, early concrete foundations had "battered footings." In this case "battered" means that one (or sometimes both) faces of the footing leans slightly so the top of the footing is not as wide as the bottom. Generally only the inside faces of footings are battered. Battered footings usually do not provide support that is aligned directly under the weight of the house; this can cause the footings to rotate, especially in clay soils. The diagram in Figure 10-5 shows how the weight of the building does not align above the support provided by the soil under the footing. The photo shows one of the signs that the footing is rotating: a gap between the bottom of the stud and the outside edge of the mudsill.

Figure 10-5 The diagram above shows a cross-section of typical foundation construction used in early concrete foundations. The building weight transfers through the studs; the soil supports the footing uniformly with a "resultant" force that acts up along the centerline of the footing. The building weight and opposing soil support are offset by the distance "e." The off-center forces cause gradual rotation of the footing. The mudsill is connected to the footing and rotates with it as a gap appears and widens at the outside edge of the stud, as shown in the close-up photo at right.

There are a few tell-tale signs of footing rotation. From the crawlspace you usually see the gap shown in Figure 10-5; other signs include leaning studs and leaning footings. Cracks where footings meet at corners indicate footing rotation if they are widest at the top. If the house has stucco on the exterior a horizontal crack will often appear along the mudsill. See Figure 10-6 for cracked stucco & footing. Sometimes a gap between the foundation and an interior concrete floor slab is mistaken as a sign of rotation—concrete shrinkage may have caused such a gap, so you can usually rule out footing rotation if you do not see other signs.

Opinions among engineers differ on when footing rotation becomes a problem and how to address it. If footing rotation continues it eventually reaches a point of "positive feed-back" where greater rotation magnifies the forces causing the rotation, resulting in rapid failure. When this point is reached depends on the footing height and cripple wall height, with shorter footings and taller cripple walls being preferred.

If a footing has rotated more than about five degrees (if the mudsill slopes more than about ½" across its typical 6-inch width) then it should be replaced. Replacement is optional in other cases, but rotation should be addressed if it is occurring. A consultation with a local soils (geotechnical) engineer is money well spent on the support system for your home. Expansive soils can cause extensive damage to houses, and soft clay soils that allow footings to rotate easily usually also have significant shrink/swell properties.

Figure 10-6 Telltale signs of footing rotation:
Left: A horizontal stucco crack along the mudsill, usually a few inches above the ground;
Right: Foundation cracks at house corners, widest at the top (as viewed from inside the crawlspace).

If you do not have expansive soil or other concerns you can stabilize existing footings with supplemental "bench piers" placed at intervals along the outside edge of the footing. If the footing is rotating partly because it was installed in poorly-compacted soil, then the bench piers must be dug down to undisturbed soil. Most contractors charge almost as much to install bench piers as to replace the footing outright; bench piers leave you with a slightly-rotated, outdated, ugly foundation that future homebuyers will question, all at only slight savings over a new footing.

In my opinion, slightly-rotated footings are unlikely to cause an otherwise well-constructed earthquake retrofit system to fail, as long as there are not other significant foundation problems. However, rotation usually indicates a substandard foundation that you would be wise to replace.

10.5 *Capped footings*

Many older homes were built before codes required at least six inches of separation between ground level and the mudsill. Some homes originally constructed with adequate soil-to-wood separation had landscaping work that resulted in higher ground level against the house. One solution to this issue is to temporarily support the house, trim off the bottom of the cripple wall, and pour a concrete "cap" on top of the existing footing. Figure 10-7 shows a cap installed in 1924 on a footing that was originally placed in 1904.

Figure 10-7   Newer concrete cap on an ancient footing.

If your foundation is capped you can still attach retrofit hardware to it, assuming there are no other problems such as footing rotation or deteriorating concrete. For details refer back to Section 6.5.5

Capped footings exist only because the original foundation was not constructed high enough. They were the cheap solution at the time, and additional expenses for reinforcing bars were often spared. Some capped footings are constructed carefully; some of them are begging to be replaced with a contemporary foundation system.

10.6 Saddled footings

A footing "saddle" wraps over both sides and the top of the original footing. Saddles may be constructed for the same reason as a cap, or to repair a deteriorated footing. Unfortunately, a saddled foundation can appear as an entirely new foundation. Attentive home inspectors often find old brick hidden within what appears to be a solid concrete foundation.

Saddles make it more difficult to determine the condition of the underlying footing. As with capped footings, saddles can be incorporated into an earthquake retrofit if other problems are not substantial—and as with capped footings, a saddled footing may be ripe for complete replacement.

10.7 Footing "curbs"

Curbs are another way to address the concern of inadequate separation between the ground and the wooden portion of a structure. In my opinion this is a band-aid solution at best, and a scam at worst. Concrete curbs use the wood-framed cripple wall of your house as one side of the concrete form. Usually sheet-metal flashing is placed against the outside of the existing wall. Concrete is placed between the wall and a temporary form board to create a four- to six-inch thick curb, high enough to provide "adequate" separation between wood and soil. The problem with this system is it provides an almost completely concealed pathway for termites to travel from the ground up into the wood wall structure.

The photos in Figure 10-8 through Figure 10-10 show a foundation curb as seen from the outside, from inside the crawlspace, and also a case where the siding has been removed.

Figure 10-8 Foundation "curb" at left side of photo wraps around the uphill side of this house.	Figure 10-9 Shiny metal flashing seen from the inside of the crawlspace of house in photo at left.	Figure 10-10 Siding was removed for repairs, revealing cripple studs extending below ground level.

Curbs have no direct effect on an earthquake retrofit, though they may lead to termite infestation that requires extensive repairs before you have solid framing to which you can connect your retrofit.

10.8 "Bathtub" basements

Many a crawlspace was turned into a basement by digging down to create additional headroom. Digging straight down next to your house's foundation can lead to serious loss of support. Exactly how close you can safely dig to an existing foundation depends on many factors. If you want to excavate a basement, consulting with a soils engineer is a good place to start.

Many basement excavations have very slim margins of safety. What looks like a sturdy underpinning for the original foundation may just be a thin layer of concrete—in essence, not much more than plastered dirt.

10.9 Brick foundations

Common lore seems to indicate that brick foundations are completely inadequate for supporting a house. Oddly, some of the people making this claim will turn right around and tell you that the brick chimney survived through the last earthquake (centered 60 miles away) and is just fine.

While there is lively disagreement about the adequacy of brick foundations to support retrofit connections, I am not aware of brick *foundation* failures that lead to collapse of a wood structure. Extra attention is needed at connections to brick, as indicated in Figure 10-11.

Figure 10-11 Additional anchors compensate for relatively soft brick in this foundation. Replacing the foundation was not an option within financial reach for the owner.

If existing brick masonry is strong enough it can still be used to support *brick* buildings. Wood-framed buildings are much lighter, and the sliding (shear) stress in a brick footing works out to be around one or two pounds per square inch. This is very low stress, and if the mortar is still in solid condition (unlike the chimney mortar shown in Figure 12-2) then it would probably function adequately in an earthquake. Note that mortar deteriorates as it reacts with moisture and other airborne compounds, so it is almost always softer where exposed.

As with concrete, the question becomes one of projecting the future lifespan of the foundation. If you can afford to, replace the foundation now if it appears to have only a couple of decades of strength left in it—you'll sleep better.

10.10 Efflorescence

Concrete and masonry are porous materials and allow moisture to pass through them. Moisture can carry dissolved salts and other minerals as it moves through the concrete or masonry footings. As the moisture evaporates from the surface of the footing the salts are left on the surface. Over time the salt deposits build up a visible layer on the footing (usually appearing as white patterns). Figure 10-12 shows heavy efflorescence. If the moisture source is eliminated the efflorescence stops forming. (In California the moisture source is often excessive landscape irrigation—although in most of California, *any* landscape irrigation should be considered "excessive.") Sometimes you even find efflorescence on the surface of the ground under a house.

Figure 10-12 The white crystalline deposits on the upper half of this foundation are called "efflorescence" and resulted from moisture moving through the concrete. Over long periods of time, efflorescence can damage the surface of the foundation material.

Over very long time periods, efflorescence can damage the surface of the footing. As the salts crystallize, the growing crystals exert enormous pressure on the surrounding concrete or masonry matrix. This occurs only at the outermost layer of the footing. One way to protect from this is to "parge" the foundation, meaning to coat it with a thin layer of cement plaster. The parging becomes a sacrificial layer that must be replaced periodically.

On the east coast, "rising damp" is caused by the same conditions that lead to efflorescence. In California's relatively dry climate the moisture may evaporate fast enough that the surface of the foundation never even appears to be damp.

The bigger concern is usually moisture related problems such as mold or increased likelihood of termite activity. I have rarely seen efflorescence as the primary cause leading to substantial foundation deterioration.

10.11 Termite and dry-rot repairs

You cannot attach plywood bracing panels and other retrofit strengthening measures to rotted or termite-infested wood and gain any benefit. Retrofit work is often done in conjunction with the sale of a house, when termite or other damage is often repaired.

Termite, rot, and other "wood-destroying organisms" are addressed in California by the Structural Pest Control Board. This board, and the contractors it licenses, apparently know almost nothing about how earthquake forces affect buildings. Sometimes I wonder if they understand how gravity affects buildings.

Pest control contractors are concerned only with removing damaged wood and replacing it with an equivalent amount of new wood. No consideration is given to the integrity of the structure. If the center six feet of a 16-foot board is damaged, you end up with five feet of

original board at each end of a new six-foot board. If the original board was intended to tie two segments of the building together it will likely need to be supplemented after the pest repair work is done. Figure 10-13 and Figure 10-14 show connections that would have warranted a severe flogging for a carpenter, yet pest control firms routinely make such "repairs".

If you buy a house that needs "pest repairs" done, make sure you have any earthquake retrofit work done *afterward,* so that the retrofit contractor can patch together what the pest control contractor butchered.

WARNING: Pest control contractors typically love to use preservative-treated lumber—Section 8.5 explains why you should insist on borate-treated lumber for any treated repairs.

Figure 10-13 Termite-damaged material on the left side of the shiny metal plate was removed and replaced with new wood. Several other joists were spliced in the same manner at the center of their spans (where bending stresses are greatest) with plates that have no structural rating from their manufacturer.

Figure 10-14 All three members of the triple stud in the center of the photo were cut and spliced at mid-height by a pest repair contractor. The 1x4 ledger covering two of the splices provides little stability for this connection, which would never be allowed in new construction.

Chapter 11 Stucco Can Resist Earthquakes in Some Cases

First let me state that stucco would never be my first choice for an earthquake resisting system. Since stucco is the main earthquake resisting element (even if only by accident) on millions of houses, it deserves discussion.

Stucco is the common term for "Portland cement plaster." This section addresses generic three-coat stucco that uses wire reinforcement; some proprietary "one-coat" or similar systems have their own specific installation requirements. As discussed later, even the "generic" installations were often installed incorrectly; a proprietary system with specific installation requirements may have suffered even more.

Stucco itself is not waterproof at all, so it must be applied over a "weather-resistive barrier," or **WRB** (sometimes called a *water*-resistive barrier). Stucco can be extremely durable—in fact it usually outlives the tar-paper often used underneath it as the WRB.

Reinforced, properly installed stucco can be very effective in resisting earthquake forces. Unfortunately, installation methods were not always appropriate for the materials used. Also decades of weather exposure (or in recent construction, attack by PT lumber) can degrade essential connections between the stucco and the structural wood framing.

Stucco installed over foam insulation board (common for some "one-coat" systems) is unlikely to have appreciable strength in an earthquake. If the WRB is visible on the interior face of the garage (or elsewhere) you can usually find a damaged spot where the foam or stucco shows through. If you are lucky enough to have an intact WRB, you can rap on it with your knuckles—if it feels soft, you probably have a layer of foam under the stucco.

Woven hexagonal mesh is the most common reinforcing material for stucco; welded-wire mesh is much less common in California. The mesh appears similar to "chicken wire," but is thicker and has a heavier zinc coating on it to better prevent corrosion. Figure 11-1 shows hexagonal stucco mesh installed over WRB.

Figure 11-1 Stucco mesh installed over weather-resistive barrier, ready for first coat of plaster.

Stucco installation methods have changed through the decades, not always for the better.

11.1 Pre-World War II practices

Two stucco application methods were used in the SF Bay Area through the 1930's. One (hopefully rare) method was essentially a non-structural skim-coat of cement plaster over wood sheathing with no WRB; the other method used wire mesh reinforcement that provides strength to resist earthquakes.

11.1.1 Unreinforced stucco over grooved sheathing

Figure 11-2 shows failing stucco on the exterior of an early 20th century house in Oakland. The stucco is applied over 1x6 redwood sheathing (called "Berkeley lath" by some) that was grooved to key the stucco to the surface (see Figure 11-3).

Anyone who has owned or worked on lath-and-plaster walls has probably experienced sheets of plaster falling off once the keys broke at the surface of the lath. Stucco with no wire reinforcing will fail in the same way, leaving only horizontal sheathing to resist lateral forces. The added danger is falling slabs of plaster.

Figure 11-2 Early stucco backing (called "Berkeley lath" locally). Grooves sawn into the redwood boards held the stucco in place—for a while. Stucco in the photo above peeled away from the grooved boards. With luck this method did not spread much beyond the East Bay.
Photo: Brian Rood, GGASHI.

Figure 11-3 Ends of grooved boards exposed at a wall opening. Note that no WRB could be installed between the stucco and grooved siding.

According to Bay Area home inspectors this method of stucco application phased out in the 1920's. From the interior, this system looks just like redwood sheathing (although some stucco may have squeezed through between boards). Identification is easy if you can see the end of the sheathing boards. From the exterior you might see horizontal cracking at even spacing.

Any sort of WRB is impossible to install with this stucco application method, as it would cover the grooves in the sheathing.

Chapter 11 Stucco Can Resist Earthquakes in Some Cases

First let me state that stucco would never be my first choice for an earthquake resisting system. Since stucco is the main earthquake resisting element (even if only by accident) on millions of houses, it deserves discussion.

Stucco is the common term for "Portland cement plaster." This section addresses generic three-coat stucco that uses wire reinforcement; some proprietary "one-coat" or similar systems have their own specific installation requirements. As discussed later, even the "generic" installations were often installed incorrectly; a proprietary system with specific installation requirements may have suffered even more.

Stucco itself is not waterproof at all, so it must be applied over a "weather-resistive barrier," or **WRB** (sometimes called a *water*-resistive barrier). Stucco can be extremely durable—in fact it usually outlives the tar-paper often used underneath it as the WRB.

Reinforced, properly installed stucco can be very effective in resisting earthquake forces. Unfortunately, installation methods were not always appropriate for the materials used. Also decades of weather exposure (or in recent construction, attack by PT lumber) can degrade essential connections between the stucco and the structural wood framing.

Stucco installed over foam insulation board (common for some "one-coat" systems) is unlikely to have appreciable strength in an earthquake. If the WRB is visible on the interior face of the garage (or elsewhere) you can usually find a damaged spot where the foam or stucco shows through. If you are lucky enough to have an intact WRB, you can rap on it with your knuckles—if it feels soft, you probably have a layer of foam under the stucco.

Woven hexagonal mesh is the most common reinforcing material for stucco; welded-wire mesh is much less common in California. The mesh appears similar to "chicken wire," but is thicker and has a heavier zinc coating on it to better prevent corrosion. Figure 11-1 shows hexagonal stucco mesh installed over WRB.

Stucco installation methods have changed through the decades, not always for the better.

11.1 Pre-World War II practices

Two stucco application methods were used in the SF Bay Area through the 1930's. One (hopefully rare) method was

Figure 11-1 Stucco mesh installed over weather-resistive barrier, ready for first coat of plaster.

essentially a non-structural skim-coat of cement plaster over wood sheathing with no WRB; the other method used wire mesh reinforcement that provides strength to resist earthquakes.

11.1.1 Unreinforced stucco over grooved sheathing

Figure 11-2 shows failing stucco on the exterior of an early 20th century house in Oakland. The stucco is applied over 1x6 redwood sheathing (called "Berkeley lath" by some) that was grooved to key the stucco to the surface (see Figure 11-3).

Anyone who has owned or worked on lath-and-plaster walls has probably experienced sheets of plaster falling off once the keys broke at the surface of the lath. Stucco with no wire reinforcing will fail in the same way, leaving only horizontal sheathing to resist lateral forces. The added danger is falling slabs of plaster.

Figure 11-2 Early stucco backing (called "Berkeley lath" locally). Grooves sawn into the redwood boards held the stucco in place—for a while. Stucco in the photo above peeled away from the grooved boards. With luck this method did not spread much beyond the East Bay.
Photo: Brian Rood, GGASHI.

Figure 11-3 Ends of grooved boards exposed at a wall opening. Note that no WRB could be installed between the stucco and grooved siding.

According to Bay Area home inspectors this method of stucco application phased out in the 1920's. From the interior, this system looks just like redwood sheathing (although some stucco may have squeezed through between boards). Identification is easy if you can see the end of the sheathing boards. From the exterior you might see horizontal cracking at even spacing.

Any sort of WRB is impossible to install with this stucco application method, as it would cover the grooves in the sheathing.

11.1.2 Reinforced stucco

Up until the mid-1940's most stucco was applied over lumber sheathing. A layer of tarpaper was applied to the sheathing; this was followed with a layer of reinforcing mesh that was installed with "furring nails" or other special fasteners (see Figure 11-4) that held the mesh slightly away from the tarpaper.

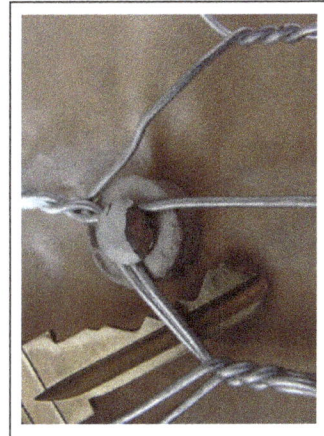

Figure 11-4 "Furring nails" come with cardboard spacers that hold stucco mesh away from the WRB.

Except for deterioration of the tarpaper behind older stucco, and subsequent fastener corrosion, decay, or termite infestation, older stucco can resist earthquake forces quite well. Most tarpaper disintegrates after several decades, though. Older stucco is seldom reliable to protect you without supplementing the anchorage as described in Section 11.6.4. Sheathing underneath stucco is often severely deteriorated—in this case even supplemental stucco connections may not provide sufficient earthquake resistance. Removal of the stucco and replacement with new exterior sheathing and stucco or siding is the best way to assure good performance. Houses don't last forever.

11.2 Line-Wire Stucco

In the 1940's or so, someone thought "why are we installing wood sheathing behind what is essentially reinforced concrete cladding?" The response was: without some sort of backing, the tarpaper would get pushed inward and tear when the first coat of wet stucco was trowelled on. The solution to this was to install "line wires" before the tarpaper, spaced about 6 inches apart and stretched tightly. Figure 11-5 shows line-wire stucco seen from inside a garage. The wires were needed only until the first coat of stucco hardened—which is good, because they stretched and slackened quickly. Wires exposed in unfinished garages, crawlspaces, or attics are loose, which has no effect on the stucco.

Figure 11-5 "Line-wire" stucco installation viewed from inside a garage; wires stretched across the studs keep the tarpaper from getting pushed in when stucco is troweled onto the wall from the far side.
Photo: David Heilig, GGASHI

The installation procedure for line-wire stucco was:

1. Wires installed and stretched tight
2. Tarpaper applied over entire wall, starting at bottom, with layers lapped to shed water toward outside of building

3. Stucco mesh installed. Mesh was usually manufactured in 39-inch wide rolls, and subsequent courses of mesh were lapped an inch or two at edges and ends. Furring nails held the mesh slightly away from the tarpaper to provide better embedment in the stucco.
4. First coat of stucco applied; the freshly-applied stucco was raked with a "comb" to create a keyed surface that created a positive bond for the second coat. The raking of first layer of stucco led to the common term of "scratch coat." The scratch coat is supposed to be 3/8" thick; the stucco mesh is (mostly) embedded in this layer.
5. The second and third layers of stucco are applied (called "brown coat" and "color coat," respectively) to give a theoretical total thickness of 7/8." The total thickness varies; the biggest factor affecting thickness is how much the line wires allow the tarpaper to bulge into the stud cavities.

11.3 "Self-furred" Stucco

A few more "improvements" in stucco application occurred in the 1950's. These were implemented in rapid succession or at the same time; most stucco used in residential construction up through the 1990's was applied using these methods.

11.3.1 Elimination of line-wires

One labor-saving change was to eliminate the line-wires; apparently the support they provided to the WRB was not needed, or some other explanation. This change probably had no significant effect on the structural performance of the system. Greater variance in the thickness of the stucco could lead to more distinct crack patterns at the thin areas over studs.

11.3.2 Combining mesh and WRB— when "new and improved" *isn't* improved

The worst improvement in the stucco industry was to manufacture mesh with WRB attached, so that workers could unroll the WRB and mesh in the same operation. *When installed correctly* this system was just as effective as line-wire application where the WRB was installed before any mesh. Unfortunately this system using "paper-backed lath" was installed incorrectly with great frequency.

Figure 11-6 shows the proper installation method for paper-backed lath, using the following sequence:

1. Install first course of lath

2. Bend stucco mesh out at top, so that the WRB in the next layer of lath will not cover it

3. Install second course of lath so that it overlaps *only the paper layer* in the first course (left photo)

4. Bend stucco mesh from Step 2 back over the mesh in the second course, so the mesh layers are in contact and overlapping (right photo)

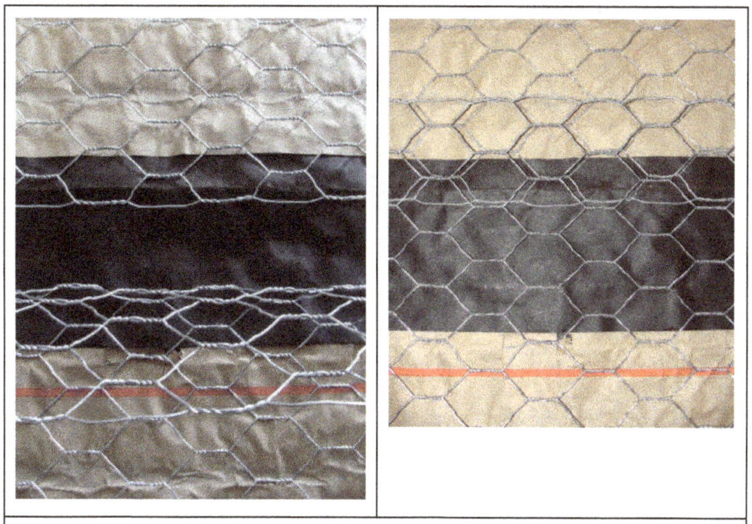

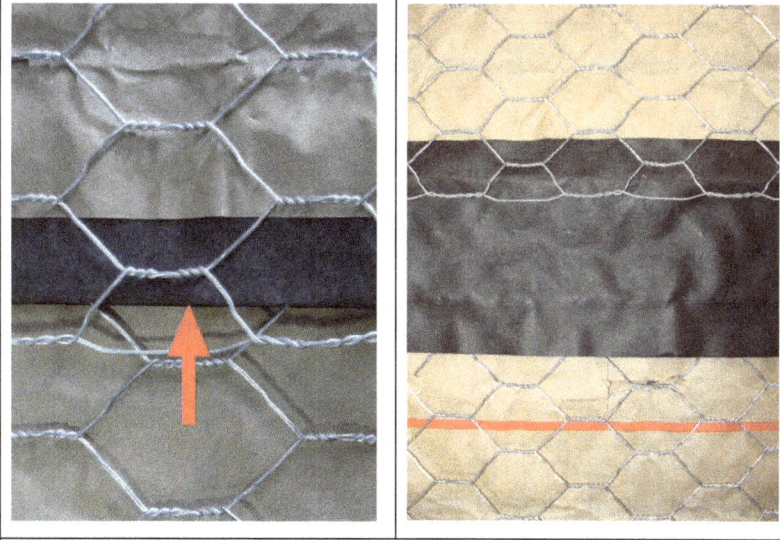

Figure 11-6 Proper installation of "self-furred lath:" *Left:* Mesh for lower course is bent back, so WRB of upper course will lap onto WRB of lower course. *Right:* Mesh laps at least two inches between courses.

Mis-Installation Steps:

Failure to follow the lath manufacturer's instructions resulted in improper installation of paper-backed lath in a significant portion of houses (perhaps 30% or more).

1. Install first course of lath.

2. Install second course of lath with paper backing overlapping the mesh of the first course.

   Figure 11-7 shows two cases of defective self-furred lath installations. As you can see, this results in a break in the continuity of the reinforcing mesh. It also creates a sudden change in the stucco thickness that may result in formation of horizontal cracks.

Figure 11-7 Incorrect installation of self-furred lath: *Left:* Subtle, but wrong; mesh from lower lath disappears under the black strip of paper behind the upper lath; mesh laps only about ½" *Right:* Blatantly wrong; paper flap covers mesh, creating 3-inch wide band with no mesh reinforcing at all, let alone a lap-splice.

In addition to the above, most lath is made with a top and a bottom edge; if you install it upside down, the WRB and mesh will not overlap properly no matter what else you do.

Tell-tale mis-installation signs:

The back side of stucco is often exposed in garages; if not, you may be able to view it in the crawlspace (if present) or from inside the attic at gable-end walls.

Figure 11-8 shows the back side of two different types of defective installations; in the photo where a two-inch band of mesh is visible, the lath was installed upside-down in addition to being lapped incorrectly. In more stealthy, but still defective, installations the mesh is sandwiched between the layers of WRB at laps. You can separate the paper layers enough to glimpse the mesh, or feel it behind the paper at the laps.

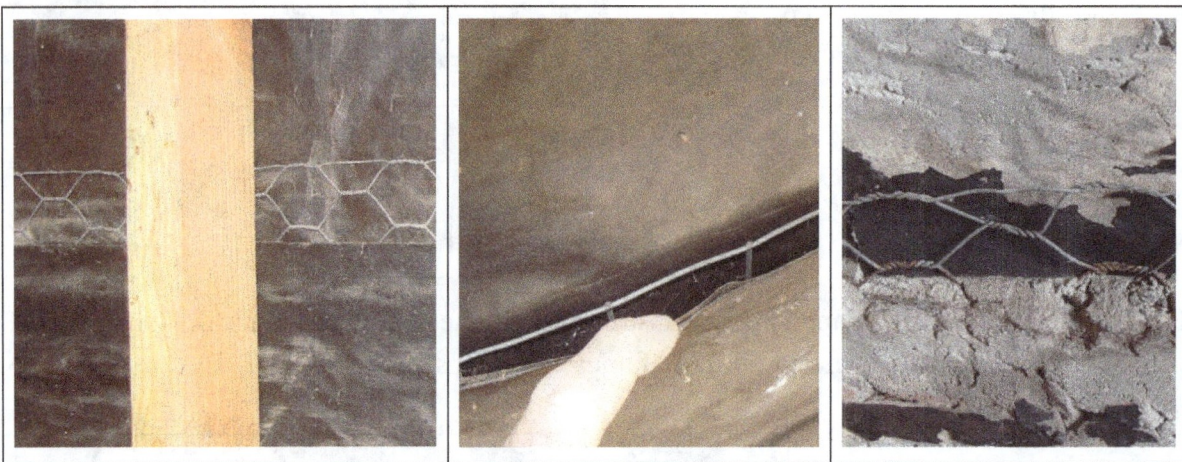

Figure 11-8 **Defective stucco installations** seen from interior side of wall;
Left: Exposed band of mesh indicates the lath was installed upside-down as well as lapped incorrectly; *Center:* Rectangular mesh peeking out between paper layers;
Right: Paper layers disintegrated with age, showing mesh emerging from stucco.

Horizontal exterior cracks in the stucco, especially when evenly-spaced about three feet apart, are a fairly clear indication of a defective installation.

11.3.3 Poor location of mesh within the stucco

The "self-furring property" is provided by forming rows of indentations along the wire mesh. The indents hold the mesh away from the paper and make the system "self-furring." The indents are shown in Figure 11-9; this is a very subtle degree of furring, and will barely hold the mesh away from the paper. The ideal place for the mesh is close to the center of the three-coat total build-up; furring nails provide a much more positive way to locate the mesh where it provides greatest reinforcing for the stucco.

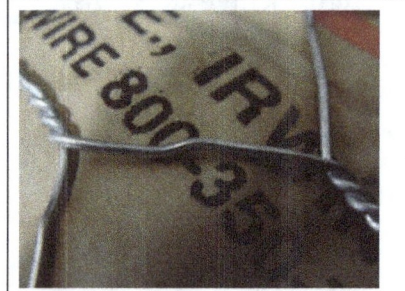

Figure 11-9 Slight bend in wire at center of photo holds it away from the WRB.

11.3.4 Inadequate fastening of lath

Paper-backed lath is often applied with heavy-gauge staples. The staples are *supposed* to be placed where the self-furring indents occur. Since the wall framing is no longer visible once the stucco lath is rolled out, it may be hard to tell whether both staple legs penetrate the framing. Poor connection to framing allows for large slabs of stucco to detach during an earthquake. In addition, staples allow more slippage and movement than furring nails.

11.3.5 Ongoing mis-installation of self-furred, paper-backed lath

When improper installation methods have prevailed for decades, everyone involved in construction overlooks them. Improper stucco lath installation is still common today. Recently stucco is used less and less as part of the structural system that resists earthquakes. It is usually installed over plywood or OSB sheathing or shear walls. Improper mesh laps and poor lath connection are still too common, but at least they are less likely to compromise a house's primary means of resisting earthquakes.

11.4 Weep screeds

Beginning in the early 1960's, galvanized sheet metal "weep screeds" at the bottom of the stuccoed wall became a code requirement. Instead of running the stucco all the way down to ground level, the weep screed terminated it just below the mudsill. Any water that penetrated the stucco traveled down the outside face of the WRB; once it reached the weep screed it exited through weep holes at the bottom of the screed. Terminating the stucco above ground level helped to prevent termites from travelling undetected between the stucco and the foundation, going up from the ground to feast on the house framing.

Weep screeds are not a structural problem in and of themselves, but may have led to stucco mesh installation problems. When self-furring stucco lath is applied with staples over the weep screed, the staple legs often just bend rather than penetrating the sheet metal. The connection along the mudsill is crucial to a complete load path; bent staples that don't penetrate the weep screed leave a serious gap in the load path.

11.5 Stucco identification

Some stucco application methods can provide ample strength to resist earthquakes, provided that connections are still intact. Other methods should be treated with suspicion almost universally. It is important to properly identify what sort of installation you have. The following signs may help pinpoint installation methods used, especially considering the age of a building.

Unreinforced stucco over grooved wood sheathing may have evenly spaced horizontal cracks as described in Section 11.1.1. There may also be bulging areas where the "keys" that hold the stucco to the wood have cracked and the stucco is beginning to pull away from the wall; this delamination is a danger sign that warrants prompt attention.

Reinforced stucco applied over solid lumber sheathing will likely have cracks beginning at the corners of window and door openings. These cracks usually slope at 45 degrees. There may be other more or less random cracks throughout wall areas away from openings.

Line-wire stucco often has vertical hair-line cracks that occur over studs. (Even with wires providing support, the tarpaper WRB deflects inward when stucco is troweled onto the wall; this results in thicker stucco between studs—the stucco cracks where it is thinnest: at the studs.)

Self-furring stucco usually also has vertical cracks for the same reason described for line-wire. If there are also horizontal cracks spaced about three feet apart, or that extend for more than a few feet, it is likely that the paper-backed mesh was not lapped correctly.

Weep screeds indicate construction since about 1963 or later.

Figure 11-10 Vertical hair-line cracks often occur over wall studs in line-wire stucco. Photo: Chris Shupp, Silicon Valley ASHI-CREIA

Stucco applied over sheathing will not vary in thickness as much as line-wire or un-backed self-furring stucco, and will not have as many vertical cracks. Sheathing varies greatly in structural integrity, ranging from fiber-board to plywood.

11.6 Retrofitting stucco

As with any structural system, there are many variables that will affect whether and how to retrofit a stucco-clad building. Definitive advice can come only from a competent professional based on a site investigation of your particular circumstances. The following is provided only as general guidance.

11.6.1 Special considerations regarding stucco

As discussed in Section 3.5.4, stucco is a very stiff structural material when compared to wood framing and plywood bracing. If we use stucco to resist earthquakes we need to design for much higher forces than we do for plywood shear walls. Even if plywood is used as retrofit cripple wall bracing, I recommend increasing mudsill connections to resist at least twice as much force as calculated for plywood.. Even if the stucco fails and falls off the building, it will not do so until after the initial earthquake jolt; we do not want the initial jolt to overload the mudsill connections.

11.6.2 Findings after the 1994 Northridge Earthquake

A task force convened by the City of Los Angeles studied earthquake-related damage and released a preliminary report in May, 1994. The task force made recommendations for code changes affecting *new* construction, which include:

- Prohibit using staples to attach self-furring lath
- Require that wire lath is attached to framing members with furring nails to provide 3/8" embedment of lath in the stucco
- Limit building height to one story when stucco is the only bracing material; stucco may not be used to brace cripple walls.
- Reduce the allowable strength of stucco to half of value allowed in prior Codes.

11.6.3 Conditions that warrant concern

Rebuilding millions of homes simply because they have stucco cladding would be insane. Assuming that the stucco will contribute nothing to a house's ability to resist earthquakes is not reasonable in many cases. I recommend repairing stucco or providing some supplementary structural system under the following circumstances:

- Unreinforced stucco on grooved boards, especially if it is delaminating
- "One-coat" stucco systems
- Improperly lapped mesh in self-furred stucco
- Stucco applied over fiber-board
- Weep screeds were used (especially if there is any way to tell that staples were used to fasten the lath, as they typically could not penetrate the weep screed and connect it adequately)
- Single-story walls where the width of stucco is less than 1/3 of the total wall length (less than ½ of total width for tile roofs)
- Two-story construction (or cripple walls below one-story construction), where the stucco width is less than 2/3 of total lower-floor wall width (less than 3/4 total width for tile roofs).
- Any installation where water damage may have occurred.

As usual, you would best seek the advice of an experienced engineer regarding the above.

11.6.4 Potential retrofit method

The late structural engineer Ben Schmid conceived, developed, and tested a simple retrofit method for stucco. His system uses special nails to supplement the connection along the mudsill and possibly other "boundary" members in the stucco wall system. The key to the system are heavy-gauge hot-dip galvanized nails, similar in shape to duplex-head nails, that are installed in holes drilled in the stucco. A special drill forms a small pocket in the stucco for the nail; after nails are installed, the pockets are filled with stucco to patch the hole and provide the connection between the original mudsill and stucco.

To my knowledge Mr. Schmid's system has not yet gone into mass production. I hope that that will change with enough interest from the retrofit community.

11.7 Further information on installing NEW stucco

The Technical Services Information Bureau (TSIB) has very detailed information on stucco installation. According to their website, the "TSIB is a not-for-profit Standards Setting Organization (SSO) that provides educational, technical and on occasions, inspection services to the wall and ceiling industry. The TSIB is a consensus bureau with input from code authorities, designers, contractors, manufacturers and dealers."

For current best-practices, see http://www.tsib.org/pdf/plaster-assemblies-chapter-02-plaster-substrates-lath.pdf

Chapter 12 Other Earthquake Hazards

12.1 Masonry chimneys are killers

I admit—I love the look of well-crafted masonry. But if they are above waist level, masonry chimneys are too dangerous in earthquakes to justify keeping. Brick patios, benches, planters, and walkways can be very charming ways to recycle your chimney.

The Modified Mercalli Intensity (MMI) scale is the preferred way to measure earthquake intensity. It relates directly to how humans and structures react to a quake. Earthquake intensity of VII on the MMI scale corresponds to "*some chimneys broken or heavily damaged.*" (This corresponds to about 5.5 to 6.0 on the Richter scale.) If an earthquake caused occasional "*fall of chimneys*," it was around VIII on the MMI scale. (Note: There are *four* intensity levels that are higher than the level that knocks chimneys over!). I ask my clients, "Why would you want to live next to an earthquake indicator?"

Figure 12-1 Many people look at a chimney and think "home." Engineers and home inspectors think "danger." Geologists think "earthquake sensor."

12.1.1 The pure economic approach

The 2006 IEBC Commentary suggests that removing a chimney and filling in the empty space left in your house costs about the same as repairing the typical damage it would suffer in an earthquake. Given the choice of spending money now to remove the chimney, or waiting for an earthquake that may never come during your lifetime, it's tempting to not spend the money. But if your chimney falls down and kills you or injures someone, you made the wrong choice.

12.1.2 Mortar disintegration

Until 1930 or so, most mortar was made with lime instead of the more modern Portland cement. Lime mortar gains strength for a while after construction; at some point its strength peaks, then it begins to deteriorate from exposure to moisture, acid rain, small earthquakes, death metal music, etc. In my experience, all lime mortar used chimneys reached peak strength and is now in significant decline. Often the disintegrating mortar is hidden in an attic, crawlspace, or fireplace-surround, or perhaps concealed under paint or stucco; Figure 12-2 shows some examples of chimney deterioration and damage.

Mortar vanished from brick joints in basement. There is about 20' of chimney above this point.	Exterior brick joints losing enough mortar to approach imminent collapse.	Mortar loss about 8 feet from bottom of a two-story chimney

Fireplace under demolition—the single layer of brick is typical of older fireplaces, leaving a hollow, very weak structure.	Many chimneys have very shallow footings (as little as 8 inches deep) and may already be leaning. Note the gap between the chimney and the siding.	Weak mortar joints shaken apart by prior earthquakes. Photo by Roger Robinson

Figure 12-2 Various chimney conditions. Inspectors find conditions like the ones shown above with alarming frequency. When I started my engineering career in 1990, a simple test of mortar strength was whether you could scrape it away with a house key. Most California chimneys are now 20 to 50 percent older now, and I often find mortar that I can brush away with my fingertips.

12.1.3 Chimneys of death, or at least destruction

Figure 12-3 shows chimney damage caused by various earthquakes. After the Napa earthquake of 2014 I will recommend chimney removal even more strongly. When replacing roofs, some roofing contractors would rather remove your chimney down to the roofline at no charge than worry about waterproofing the joint between the brick and the roofing.

For a fun video showing two people pulling an old chimney over from what they *thought* was a safe distance, see: http://www.youtube.com/watch?v=XgvhANUF1kQ&feature=youtu.be

This fireplace collapsed during the August 24, 2014, Napa CA quake and seriously injured a boy having a sleep-over.	A gentle earthquake broke this chimney where the brace connects to it.	If a chimney wants to fall, those steel straps connecting it to your roof framing will not do much to stop it.

Figure 12-3 Damaged chimneys; most areas on the west coast can expect earthquakes that will produce much more intense shaking than the quakes that caused these failures.

12.1.4 Bracing is not effective

Many building owners had braces installed on their chimneys. These may help keep the chimney from falling in moderate earthquakes, but are unlikely to help in larger quakes. One problem is the earthquake loads associated with chimney can easily double the force on the surrounding building (chimneys weigh a *lot*). Another problem is that a brick chimney is much more rigid than a wood-framed house—initially the chimney will brace the

Chimney Braces & Your Insurance Company

Chimney braces are so common that the public perceives them to be more effective than they really are. In some cases, insurance companies support this myth by requiring chimney bracing, with the implication that a braced chimney is "safe." *(continued next page)*

house, not the other way around. By bracing the chimney you are essentially assuring that it will break; what you hope is the braces (and all the connections in the load path from the braces to the foundation) are strong enough to hold the column of newly-loosened bricks in place during the shaking.

Strengthening a house enough to brace 10 tons of bricks shaking around is a very complicated and expensive undertaking, though. Running a couple of braces from a steel bracket around the chimney down to the roof is only the beginning; if you stop there, you have simply installed warning devices: if you are not speared when the braces punch through your ceiling during a quake, you will know that 10,000 pounds of brick is about to follow.

12.1.5 Change your fireplace to burn natural gas or propane

For the three nights a year when you want a cozy fire, install a gas-burning insert in the fireplace with tight-fitting glass doors and a metal vent system that replaces the chimney.

12.1.6 So, you really need centuries-old heating technology?

Wood-burning fireplaces cause air pollution, create drafts in your house that waste energy, and are expensive to operate. But if you really think a wasteful, polluting, messy heating appliance is for you, the following document shows how to replace the chimney above your fireplace with metal flue pipe: http://ladbs.org/LADBSWeb/LADBS_Forms/InformationBulletins/IB-P-BC2008-070EQDamagedChimney.pdf

> As I was taking photos of this very tasteful replacement of a Berkeley chimney, a curious neighbor struck up a conversation. When I explained myself, he excitedly told me how many chimneys had fallen or suffered serious damage in the Loma Prieta earthquake, which was centered 60 miles away. The Hayward Fault will shake about 20 times harder when it breaks loose, toppling many chimneys that are now older and weaker.

Chimney Braces & Your Insurance Company
(continued from previous page)

Another structural engineer related a telephone conversation she had with an insurance company representative about chimney bracing requirements:

Insurance agent: *"We won't issue earthquake insurance unless the chimney is braced."*
Engineer: *"You know that bracing doesn't keep the chimney from falling, right?"*
Agent: *"Yes, but our studies show that if you brace the chimney, it is less likely to fall away from the house and damage a neighboring property. If it falls on the property next door, then we have to pay two claims instead of just one."*

Is it worth increasing the danger to your life in exchange for lower insurance premiums?

Wood-framed replacement "chimney"

Are Newer Chimneys Better?

Portland cement mortar is much more durable than lime mortar, so are new chimneys "safe?" Many failures would indicate the contrary.

After talking with contractors who specialize in chimney maintenance and repair, one GGASHI member states, "**I don't trust any chimney to have sufficient reinforcing.**"

Chimney repair specialists are trained only to address the condition of a chimney as it relates to conveying hot smoke up away from the fireplace—they typically don't even check to see if the chimney will withstand an earthquake.

The photo below shows one of many failures of "modern" chimneys.

Figure 12-4 *Left:* A chimney toppled by the 2014 Napa, CA earthquake. This chimney was built around 1980, over 30 years after codes began requiring steel reinforcing in chimneys. By then an entire generation of masons should have known about the "new" code requirements for this basic and essential reinforcing; how much do you trust *your* chimney?
Right: Floor framing crushed by the chimney where it landed.
Photos courtesy of Dan TerAvest, newly-avowed advocate against chimneys.

12.2 "Non-structural" earthquake hazards

Wood-framed structures rarely cause direct injury or death in earthquakes. In the Napa earthquake that occurred on August 24, 2014, the worst injury was from a collapsing chimney described in the previous section. A falling television set caused injuries that claimed a victim's life about 10 days after the earthquake.

This book is aimed as much at protecting your home and investment as protecting your life. Even if you strengthen your house, a falling bookshelf or wardrobe could kill you. This section is meant to bring non-structural hazards to your attention, but addressing them is far beyond the scope of this book.

Non-structural hazards cause a many injuries in earthquakes. If you live in a wood-framed house, you are more likely to be injured by the contents of your house than by the house itself, even if it is not retrofitted against earthquakes. Homeowners can eliminate many earthquake hazards themselves. Most of the following are "non-structural" earthquake hazards that do not usually require engineered solutions to correct. The Association of Bay Area Governments (http://quake.abag.ca.gov/) has information on how to address many of the following hazards:

12.2.1 Water-heater strapping

Under California state law, water heaters must be braced before a house can be sold. **Most of the time, such bracing or strapping is not installed adequately.** The Golden Gate chapter of the American Society of Home Inspectors has a short video on their website that shows how to check if water heater bracing is secure, at the following link: http://www.youtube.com/watch?v=RrUqO2HJIns

12.2.2 Gas shut-off valves

You should know where your gas supply is and how to turn it off. When I began writing this section, I tried to turn off my gas valve. I started with an 8-inch adjustable wrench—when that didn't work, a 10-inch wrench, and finally got the valve to budge with an 18-inch pipe wrench. I'm 6'-5" and weigh about 200 pounds. When was the last time you actually "exercised" your gas valve? Do it now.

Manual shut-off wrenches are fine if you are home when the quake strikes, your gas meter is accessible, it is not dark, etc. Many older homes have the gas meter inconveniently located in the crawlspace behind a door that is painted shut, as shown in Figure 12-5.

The best solution is to have a plumber install a motion activated automatic gas shut-off valve at your gas meter or propane tank. ("Excess flow" valves work only when a huge volume of gas is leaking out—far more than you need to fuel a fire.) Figure 12-6 shows one popular brand of automatic shut-off valve (two of these are visible in Figure 12-5). Gas shut-off valves do not work if piping between the valve and the street is torn apart as shown in Figure 12-7. Don't stop at installing a shut-off valve—brace you house, too.

For California state certification information, see: http://www.dgs.ca.gov/dsa/Programs/programCert/gasshutoff.aspx;
For a list of state-certified valves (as of October 27, 2008—this may not get updated) see: http://www.documents.dgs.ca.gov/dsa/gas_shutoff/List_of_Cert_ESV_rev10-27-08.pdf

12.2.3 House contents, especially tall, heavy furniture

ABAG has good information on how to secure shelves and other tall furniture such as bookshelves or file cabinets, dressers, china cabinets, heavy mirrors or paintings, and so forth. For suggested bracing methods see *http://quake.abag.ca.gov/residents/contents/*

Figure 12-5 Motion-activated automatic gas shut-off valves turn off your gas supply while you are at work, on vacation, etc.

Figure 12-6 Close-up of motion-activated automatic gas shut-off.

Figure 12-7 This gas meter experienced *too much* motion. Brace your house so it will not rip your gas supply apart. Photo by H. G. Wilshire, US Geological Survey.

12.2.4 Masonry veneer

Figure 12-8 shows some brick work that would be beautiful in Wisconsin, but not on the west coast. During an earthquake, brick veneer will either fall off your home and kill you, or stay attached and roughly double the earthquake force your home must resist.

The Northridge Earthquake was a test of flimsy, code-prescribed 22 gauge steel brick ties. They performed miserably; brick veneer peeled off brand-new mansions. On the other hand, 14-gauge ties connected to continuous horizontal wire reinforcement in mortar bed joints (as is required for school construction) performed very well when anchored to properly-designed wood framing.

Brick is not the only type of masonry. Decorative concrete block, stone, and sometimes terra cotta facings are also vulnerable.

Figure 12-8 Big bad wolves are a minor danger compared to bricks raining down on panicked occupants as they run out the door.

12.2.5 Exit stairs, decks, roofs and canopies

The main concern with exterior stairs, porch roofs, or canopies is that they will collapse and trap you inside the house. Most older homes have reasonably stable stair construction. The scariest stairs I see are usually replacement stairs, or just really high porches, as shown in Figure 12-9.

Porch roofs that are "planted" onto the front of a building with flimsy attachment can detach more easily than roofs that continue from the main house. Other dangers include connections deteriorated by moisture intrusion, and multiple layers of heavy roofing material.

12.2.6 Concrete block "fences" (privacy walls)

Popular in Southern California, these can fall and crush things. Don't be one of those things: use parts of your yard next to concrete block fences for incidental activities; think compost heaps or storage sheds, not chaise lounges or children's sandboxes.

Figure 12-9 Double jeopardy—if the exit balcony collapses, it could block the lower exit door and would make escape from the upper apartment difficult.

12.2.7 Wood-burning stoves, and stovepipes that serve them

Connect all components together and to the supporting structure securely with suitable screws or other connectors. Imagine sections of your stovepipe shaking apart and falling on the family room floor while you have a nice cozy fire burning, or imagine your wood-burning stove getting pitched across the room—maybe with a fire in it.

12.2.8 Clay appliance flues and chimneys they discharge to

Figure 12-10 shows an archaic, but still functional, flue connection from a newer furnace leading to a masonry chimney; functional until an earthquake, that is. Once the tiles shake loose the furnace becomes dangerous to use.

Millions of water-heaters and furnaces exhaust into ancient brick chimneys that are doomed to crack in small to moderate earthquakes.
A slightly-damaged chimney could allow carbon monoxide to filter into your house.

12.2.9 Heavy roof tiles

A good roofing contractor could advise you on how to keep the tiles from falling during an earthquake.

12.2.10 Roof-mounted equipment such as air-conditioning units or furnaces

This stuff is ugly already; nobody will notice if you add some ugly braces.

12.2.11 Heavy or expensive equipment

This category could include water-softeners, pressure tanks, swimming pool filters, air-conditioner condensers, evaporative coolers, propane tanks, and so forth. Broken connections to some of these could cause flooding or other secondary damage.

Dances with Wolfs (and other heavy appliances)

Heavy kitchen ranges are strangely popular. Just the *grates* from a Wolf range could cause serious injury if thrown about during an earthquake.

Cast iron wood-burning stoves are usually assumed as stable, immovable objects; not so in an earthquake! ASHI home inspector Dave Heilig recounts the following after the Loma Prieta earthquake:
We visited one house where a wood-burning stove had been installed on a platform in the living room. We found the stove on the floor just below its platform; we also found the imprint of a medallion on the upper corner of the stove deeply embedded in the dining room wall, fifteen feet away.
Imagine if a fire had been burning in the stove at the time.

Figure 12-10 This new furnace uses original clay flue tiles that lead to a chimney about 12 feet away. When the tiles shake loose in a quake, hot, toxic flue gases will discharge into the crawlspace.
Photo: Brian Cogley, GGASHI

12.2.12 Porticos, heavy trellises, & awnings

The main concern with such structures is they will trap you inside your house if they collapse.

12.2.13 Heavy objects hung from walls

Tall, heavy furniture such as bookshelves or file cabinets, dressers, china cabinets, heavy mirrors or paintings, and so forth can fall and cause injuries during earthquakes. For suggested bracing methods
see *http://quake.abag.ca.gov/residents/contents/*

12.2.14 Hazardous windows:

Non-tempered windows can break into possibly deadly pieces. Windows near beds or cribs are of particular concern. Consider installing a film to such windows so that broken pieces of glass will not scatter. Special films that will greatly strengthen the window are also available, such as 3M's "Scotch-shield."

Figure 12-11 Always check to see that totem poles are braced before entering below them.

12.2.15 Forest fires—plan your escape!

Wildfires can be pretty terrifying, and can occur in California from May or June through November—or possibly year around if the current drought is "normal" under an altered climate. Imagine an earthquake devastating many of the roads, bridges, waterlines, and buildings in your area, demanding all the emergency responders attention—and *then* having a wildfire start.

Figure 12-12 Devastating fire is a serious threat after an earthquake in California's dry, hilly terrain. Plan your escape now.

If I lived in any of the hilly, brushy or wooded areas in California, the thought of a fire following an earthquake would scare me much more than anything else. I would want to have a sturdy bicycle for each occupant on hand (with tires properly inflated at all times) and be ready to ride downhill *before* a fire started.

12.2.16 Soil stability: Landslides and Liquefaction

Liquefaction occurs when an earthquake shakes loose, saturated, sandy soils. All the sand grains momentarily lose contact with each other and turn to "quicksand"—and stop supporting your house. Determining whether liquefaction is a likely danger would require a soils investigation (possibly $5,000 to $6,000).

Landslides and liquefaction are beyond the scope of this book, and in some cases beyond hope of retrofitting to prevent them from damaging your home.

Abbreviations & Glossary

Abbreviations

AB   Anchor Bolt

ABAG   Association of Bay Area Governments (San Francisco area)

APA   American Plywood Association (now called "*APA—The Engineered Wood Association.*")

CBC   California Building Code

CEBC   California Existing Building Code

CW   Cripple Wall

(E)   Existing (often used as an abbreviation on construction plans)

FEMA   Federal Emergency Management Association (United States)

IBC   International Building Code

ICC   International Code Council

IEBC   International Existing Building Code

IRC   International Residential Code

LFRS   Lateral Force Resisting System

(N)   New (often used on construction plans)

SDPWS   Special Design Provisions for Wind and Seismic

SEAOC   Structural Engineers Association of California

SEAONC   Structural Engineers Association of Northern California

STT   Shear Transfer Tie

USGS   United States Geological Survey

UBC   Uniform Building Code

WSP   Wood Structural Panel

Glossary

Adhesive anchor   A bolt installed in a foundation with some sort of adhesive (usually epoxy, but other adhesives such as acrylic resins may be used).

Anchor   Hardware that connects the structure to the foundation.

Anchor bolt   Steel fastener used to connect sill plates to footings. (The term "bolt" technically means a headed fastener; thus, "anchor bolt" is not entirely accurate to describe typical sill anchors).

Anchor rod   Used in this book to describe hardware that connects a tie-down to the footing. "Anchor rod" is a more technically accurate term to describe a threaded fastener that connects a building's superstructure to the footing (see entry above).

Bearing plate   A heavy steel washer installed on foundation bolts to reduce splitting of the mudsill (also referred to as "plate washer" or "square washer").

Blocking   Sections of lumber installed between framing components such as studs or joists, usually perpendicular to them.

Boundary   An edge of a diaphragm; typically along eave blocking and the trusses or rafters in line with the gable-end walls for a roof diaphragm (not the fascia or barge rafters), the end and rim (band) joists for a floor diaphragm and the end-posts and top and bottom members of a shear wall.

Collector   A structural member that connects a diaphragm to a shear wall in order to gather lateral forces spread throughout the diaphragm and deliver them to the shear wall.

Crawlspace   The area below the lowest framed floor in houses with raised floors.

Cripple   In carpentry, "cripple" simply means shorter than normal. Cripple studs and cripple walls are the most common use in this book.

Cripple wall   A wall built on top of a footing, that extends up to and supports the floor framing above the crawlspace. Usually cripple-walls are only a few feet tall.

Chord   The top or bottom member of a truss; also used to describe the member along a diaphragm boundary (diaphragm chord).

Diaphragm   A large area of sheathing such as a floor, roof or shear wall (diaphragms may be vertical as well as horizontal); in order to function as a diaphragm, all edges of the sheathed area must have boundary members, and the sheathing panels must connect to these members and to intermediate framing members to transfer forces across the panel joints.

Double plate   Two horizontal framing members at the top of a wall, usually two 2x members one on top of the other, typically spliced or lapped to give continuity or tie wall sections together.

Douglas fir   The most common kind of lumber used for framing in the western US.

Drag strut Essentially the same as a collector; may have slightly varying meanings, such as a collector that functions only in compression.

Drag tie Essentially the same as a drag strut or collector, with the added possibility that it may refer to a member that transmits only force from one part of the structure to another part some distance away, without collecting any additional force along the way.

Edge When referring to a shear panel, a panel has four edges—two of these edges are "ends" and two are "sides."

End joist The last joist in a series, parallel to the nearby joists.

Epoxy anchor A concrete anchor installed in concrete with epoxy resin.

Flush Surfaces that align smoothly or evenly. A nail is "flush" with plywood when it neither protrudes from the surface nor is indented.

Footing A section of concrete foundation, especially the buried portion (may also include the "stemwall").

Foundation The structure installed in or on the ground to support a building. In most Bay Area homes, the foundation is of concrete. Many homes built before about 1910 and a few later ones have brick foundations. Concrete block masonry is less common than concrete; stone foundations are rare on the west coast.

Framing The structural "skeleton" of a building, including studs, plates, sills, joists, etc. In most houses, the framing is of lumber. Occasionally framing in newer homes is steel.

Header A horizontal member that spans across an opening in a wall (typically over a door or window).

Hold-downs Steel brackets installed at the ends of a shear panel to prevent the panel from rocking up and down on the foundation (they do not function to keep the house from "jumping off the foundation"). In newer homes, hold-downs may also be used to connect shear walls at upper floors to the foundation. The same hardware may be used to anchor decks to the main house structure, or for other connections.

Joist A horizontal wood member that supports a floor.

King stud A stud that extends up from the bottom plate to the top plate of a wall and connects to the end of a header.

Laminated Veneer Lumber (LVL): Structural members manufactured from wood veneer. Sold under trade-names of "Micro-lam", "Versa-lam" and probably others.

Lateral Describing forces that act in a horizontal direction; usually caused by wind or earthquakes.

LVL: See Laminated Veneer Lumber

Moment Frame (Also called "moment-resisting frame") A steel frame used to brace around a large opening such as a garage door.

Mudsill The board installed directly on top of the foundation, and the base to which other framing is attached. Also called the sill plate, foundation plate, or foundation sill plate.

Mudsill anchor Hardware used to connect the mudsill to the footing; can include anchor bolts, expansion anchors, adhesive anchors, threaded concrete anchors, proprietary anchors such as the RFA6 or UFP10, etc.

OSB Oriented Strand Board (sometimes called "wafer-board" or "flake-board"—particleboard is completely different from OSB); a type of wood structural panel composed of wide, thin wafers of wood bonded together with adhesive under high pressure and heat.

Pony wall Same as "Cripple Wall".

Pressure Treated Lumber Lumber that has had chemicals applied to it under pressure to resist termites, fungus, and other structural pests.

Pad An isolated concrete footing used to support a post or pier. Usually pad is at or below ground level.

Pier An isolated concrete or masonry footing used to support a post; may in turn be supported on a pad or directly on the ground. A pier rises above ground level.

Plate washer A heavy steel washer installed on foundation bolts to reduce splitting of the mudsill (also referred to as "bearing plate" or "square washer").

PT Pressure treated lumber (technically "preservative pressure-treated lumber").

Retrofit Installation of components for earthquake resistance in an existing structure.

Rim joist A joist at the perimeter of a floor framing system that is perpendicular to the other floor joists, which in turn butt into the rim joist.

Shear Forces that act in opposite directions and tend to cause structural members to slip past each other.

Shear Panel A bracing component for resisting side-to-side earthquake forces, usually consisting of plywood fastened to mudsills, top plates, and wall framing.

Shear transfer tie Generic term for a sheet metal connector used to attach floor framing members to the top of a cripple wall during an earthquake retrofit.

Sheathing Used in this guide to mean structural plywood or OSB panels applied to framing members.

Sill A horizontal framing member at the base of a wall, typically attached directly to a concrete or masonry footing.

Soft Story The building codes define a "soft story" when the structural system at any particular story has 70% or less of the **stiffness** of the structural system in the story immediately above. Unreinforced cripple walls are typically "soft stories" in wood-framed buildings. See "SWOF."

Sole plate   A horizontal framing member at the base of a wall, typically when attached to a wood-framed floor platform below (see also "Sill").

Stemwall   A concrete or masonry wall built on top a footing and extending above the ground to support wood framing.

Stud   A vertical component of a wall that supports the wall surfaces and the structure above. In older homes, studs are usually of lumber. Newer homes may have steel studs.

SWOF   Short for "*Soft, Weak, or Open Front.*" This term is used for a large category of residential, commercial, and mixed-use buildings. For single-family homes it usually refers to a portion of the building with a garage door opening or picture window that significantly reduces the strength of a particular wall line. Unreinforced cripple walls usually create soft *and* weak story conditions in wood-framed buildings. Large openings may or may not create a soft or weak story according to the strict code definitions. Depending on the location of the opening and the building configuration, large openings can create weaknesses that may lead to significant localized damage or collapse of part of the building.

Tie-down   A device used to keep the end of a shear wall from lifting up; also called "hold-down."

Trimmer stud   A stud adjacent to an opening in a wall. Trimmer studs support the header over an opening in a wall and do not extend all the way up to the top plate. Also called a jack stud.

Top Plate.   A horizontal component, usually of lumber, at the top of a framed wall; the top plate may be a single member or multiple members. See *double plate*.

Wall Plate.   (Usually called "top plate") A horizontal component, usually of lumber, at the top of a framed wall.

Weak Story.   The building codes define a "weak story" when the structural system at any particular story has 80% or less of the **strength** of the structural system in the story immediately above. Unreinforced cripple walls are typically "weak stories" in wood-framed buildings.   See "*SWOF.*"

Wood structural panel   Generic term meaning either plywood or OSB.

WRB   Weather (or water) resistive barrier; usually asphalt-impregnated felt, or "tarpaper" in older houses. New products such as Tyvek are used as WRBs.

Appendix A
"Standard Plan A" Strengthening Requirements

REINFORCEMENT SCHEDULE

	GENERAL INFORMATION		PLYWOOD BRACING	MUDSILL ANCHORAGE			FLOOR TO CRIPPLE WALL / MUDSILL CONNECTION		
CHECK THE BOX WHICH APPLIES TO YOUR HOME	TOTAL FLOOR AREA (SF) (1)	HEAVY OR LIGHT CONSTRUCTION	MINIMUM TOTAL BRACING LENGTH ALONG EACH WALL LINE	MINIMUM SILL ANCHORS ALONG EACH WALL LINE			MIN. NO. OF FLOOR FRAMING CLIPS (FFC)(3) ALONG EACH WALL LINE (4)		
				UFP10 (2)	1/2"ø BOLT	5/8"ø BOLT	NO. OF L70	NO. OF H10 (5)(6)	NO. OF L90
1-STORY REQUIREMENTS	800	Heavy	16'-0"	5	8	6	13	10	10
	800	Light	12'-0"	4	6	5	11	8	8
	1000	Heavy	17'-4"	6	9	6	15	12	12
	1000	Light	14'-8"	4	7	5	12	9	9
	1200	Heavy	20'	6	9	7	17	13	13
	1200	Light	14'-8"	5	7	5	13	10	10
	1500	Heavy	22'-8"	7	11	8	19	15	15
	1500	Light	17'-4"	5	8	6	15	11	11
	2000	Heavy	28'	8	13	9	24	18	18
	2000	Light	21'-4"	6	10	7	18	14	14
2-STORY REQUIREMENTS	1500	Heavy	17'-4"	5	8	6	15	11	11
	1500	Light	14'-8"	4	7	5	12	9	9
	1800	Heavy	24'-0"	7	12	8	21	16	16
	1800	Light	18'-8"	6	9	7	16	12	12
	2400	Heavy	29'-4"	9	14	10	25	19	19
	2400	Light	22'-8"	7	11	8	20	15	15
	3000	Heavy	N/A	N/A	N/A	N/A	N/A	N/A	N/A
	3000	Light	26'-8"	8	13	9	23	17	17

FOOTNOTES FROM TABLE:
1. See total floor area retrofit Construction Data.
2. When UFP anchors and bolts are used in a single wall line, UFP anchors maybe substituted for the number of bolts.
3. Not more than one angle per joist bay unless joists are spaced 24 inches on center. Where practicable install angles between joists above plywood braced panel locations.
4. Install L70 & L90 W/ 10d x 1½" nails (.148 x 1½").
5. H10 uses (8) 8d (.131" dia.) x 1½" into joist and (8) 8d (.131" dia.) x 1½" into top plates.
6. H10 floor framing clip should be used as an alternate only where accessibility makes the use of L70 or L90 impractical. L70 and L90 clips are preferred over H10 FFC.

④ REINFORCEMENT SCHEDULE

DEFINITIONS— "LIGHT" AND "HEAVY" CONSTRUCTION

<u>HEAVY CONSTRUCTION</u> is your home constructed using any of the following:
1. Stucco exterior wall finish
2. Heavy roofing consisting of concrete or clay tiles
   (Weighing up to 11 pounds per square foot)

<u>NOTE:</u>
Clay tile weighing more than 11 pounds per square foot may be considered on an individual basis. Check w/ your local Building Department.

<u>LIGHT CONSTRUCTION</u> is any building constructed using only the following.

ROOFING MATERIALS:
1. Wood shakes or shingles
2. Composition or asphalt shingles
3. Metal roofing (Weighing 5 pounds per square foot or less)

EXTERIOR WALL FINISHES:
1. Wood panel sheathing
2. Wood board siding
3. Similar light board siding

INTERIOR WALL FINISHES:
1. Gypsum board
2. Gypsum or plaster lath

CONNECTOR CAPACITY (Pounds)	and Connection Description
458 lbs.	L70 is 16 ga X 7" long uses (8) – 10d X 1-1/2" nails (.148 x 1 1/2")
600 lbs.	L90 is 16 ga X 9" long uses (10) – 10d X 1-1/2" nails (.148 x 1 1/2")
505 lbs.	H10 anchor uses (8) – 8d X 1-1/2" nails (.131 x 1 1/2")
820 lbs.	1/2" dia. bolt
1170 lbs.	5/8" dia. bolt
1340 lbs.	UFP10– Universal plate anchor

To download a PDF of Standard Plan A, go to: http://www.abag.ca.gov/bayarea/eqmaps/fixit/Plan%20Set%20A.pdf

In future editions of this book, this table will likely be replaced with the expanded recommendations in FEMA documents currently under development. (Look for "**Earthquake Strengthening of Cripple Walls in Wood-Frame Dwellings,**" to be released as the South Napa Earthquake Recovery Advisory *FEMA DR-4193-RA2* around June, 2015.)

Appendix B
Evaluating Existing Retrofit Installations

There is no substitute for individual analysis by an experienced engineer. In the Bay Area, a good earthquake evaluation would cost much less than one percent of a home's sale price.

The following are meant as general guidance to assess the overall quality of a retrofit. Determining whether a specific retrofit meets a particular strength level would require sampling the plywood nails, measuring lengths of plywood installed, tracing load paths, etc.

Good qualities to find in a retrofit:
The following indicate that the installer had at least some basic training regarding retrofit methods and recommended materials, made an effort at careful installation, etc.

- Anchor bolts with 3-inch square washers
- Plywood nails installed along chalk-lines that mark framing member locations
- Plywood panels cut to fit neatly
- No plywood nails over-driven
- Minimum plywood thickness of 15/32" (or decimal thickness of 0.45 inches)
- Structural I rated plywood, and/or five-ply plywood
- Hot-dipped galvanized nails
- Existing joints in cripple-wall top plates splice with steel straps or additional nails
- Blocking (where installed) fit neatly into place
- Blocking installed behind all horizontal panel joints.
- All edges of every piece of plywood supported by framing
- Bracing elements along all sides of the house (or at least "major" lengths of walls on all sides—for houses with complex floor plans this is a difficult criterion to determine)
- Collectors
- Hillside anchors
- Shear-transfer diaphragms
- Borate-treated lumber used in all instances where treated lumber is needed
- Stainless steel fasteners when connecting to ACQ, CA, CC, or ACZA treated wood

Conditions you do *not* want to find:
- Over-driven plywood nails
- Un-blocked shear panel edges
- Insufficient plywood nailing (nails spaced too widely, edges with no nailing at all, etc)
- Nails installed within less than 3/8" of plywood edges
- "Shiners" on the back side of bracing panels
- Long slots cut in shear panels without a plywood patch glued over the entire slot

- Strengthening measures not installed on all sides of house
- Oriented Strand Board, or "OSB" (though the performance of OSB is *supposed* to equal that of plywood, inspectors see many instances of failed OSB panels in the field)
- Termite or dry-rot damage
- Strap-type "hold-downs" used to connect joists to foundation with no connections to resist sliding
- Mudsill connections made, but no bracing panels installed
- Plywood not nailed along bottom edge, or plywood toe-nailed to mudsill
- Mudsill anchors installed into crumbling concrete
- Angle-irons installed to resist uplift only
- Angle-irons installed to blocking at joists
- Plywood nailing not uniform around panel
- Finish nails or roofing nails used to attach plywood

Work that may look impressive, but offers little protection:
- Tie-downs installed without bracing panels
- "FSA" or "FJA" type anchors
- Post-to-beam connections on multiple posts in crawlspace
- Tie-downs connected from posts to piers

Items that do not represent current best practices, but may be adequate:
Most of the following conditions would need to be evaluated by an experienced engineer.

- 3/8" or 7/16" thick plywood (if there is enough installed it may not be stressed to the point that it fails in an earthquake; determining how much is "enough" is an engineer's job)
- Three-layer or four-layer plywood (again, if there is enough installed to keep stresses low)
- Lack of 3" square washers at foundation anchors (when the anchors are no longer accessible because otherwise adequate plywood panels were installed over them, I do not recommend removing the panels just to add square washers)
- Use of "sheathing grade" plywood (instead of Structural I)
- Clipped-head nails; See comments in *"Wood-Framed Shear Wall Construction—an Illustrated Guide"*

Appendix C
Common Hardware Items

The following items are often used in retrofit work (if something does not appear below, it does not mean it isn't suitable for retrofit use). This is not an endorsement of any particular product.

NOTE: The following items may or may not be suitable for a particular use. Manufacturer's products change frequently. Load ratings, fastener requirements, corrosion resistance, and other factors should be verified by the designer or installer before use.

Item	Manufacturer	Catalog #
Tie-down	Simpson	"HDU" Series
	USP	"PHD" Series
"Universal" Foundation Plate	Simpson	UFP10
Flat Foundation Anchor Plate	Simpson	FAP
	KC Metals	RFP
Folded Anchor Plate	KC Metals	RFA86/88
	KC Metals	RFA136/138
	USP	SFA8
"Hurricane" Tie (Note: These are about the only hurricane ties worth using in earthquake retrofit work)	Simpson	H10A
	KC Metals	HT10A
	USP	RT16A
"Hurricane" Tie (see above) For rough lumber sizes (most pre-1930 houses)	Simpson	H10AR
	KC Metals	HT10R
	USP	RT16AR
16 gauge angle	Simpson	L90
	KC Metals	CA90
	USP	AC9
(continued)		

Item	Manufacturer	Catalog #
16 gauge angle	Simpson	L50
	KC Metals	CA50
	USP	AC5
2" wide straps	Simpson	ST6236
	KC Metals	TS36
	USP	KST234
2" wide straps	Simpson	ST6224
	KC Metals	TS24
	USP	KST224
1-1/4" wide straps	Simpson	MSTA36
	KC Metals	TSA36
	USP	MSTA36
1-1/4" wide straps	Simpson	MSTA24
	KC Metals	TSA24
	USP	MSTA24
Flat tie-plates	Simpson	LTP4
	KC Metals	FAL
	USP	MP4F
Embossed tie-plates	Simpson	LTP5
	KC Metals	FAL5
	USP	MP6F

Appendix D
Responses to Arguments Against Flush-cutting Mudsills

Advantages of the Flush-Cut Method (see 6.2.2.1)

1. Most closely matches the construction that has been tested for 60 years by APA--the Engineered Wood Association
2. Exposes the core of the mudsill, allowing you to evaluate the condition of the wood (see Figure 3-1 and Figure 6-22 for an example of concealed decay in a redwood mudsill)

Arguments against the flush-cut method

I have heard several arguments against the "flush-cut" method

1. It is impossible to cut the mudsill flush against the studs
2. Even if it was possible it would be too expensive
3. Removing part of the original structure will weaken it
4. Cutting off part of the mudsill will increase the tendency of the footing to rotate
5. Trimming material off the edge of the mudsill affects the performance of the remaining piece.
6. The remaining width leaves less room for installing mudsill anchors

Only the last argument has much merit; here's why:

1. With the proper tools, flush-cutting is reasonably easy.
2. Cost should be left for the contractor and owner to worry about. Furthermore using the flush-cut method yields a more reliable retrofit, so the cost is justified.
3. Removing *necessary* parts of the original structure would be harmful. Let's reverse the question: Would building a shear wall with mudsills, studs and plates of equal width be cause for concern? No—we do this all the time in new construction. So why would we worry about creating a mudsill that is the same width as the studs it supports?
4. Footing rotation is no more likely to occur if you remove part of the mudsill. Section 10.4 discusses footing rotation. Figure 10-5 shows the offset forces that lead to footing rotation; the only thing shown in the diagram that will affect footing rotation is the distance between the building weight and the soil supporting the footing. The photo in the figure shows evidence of a rotating footing even with a full-width original mudsill. Providing a mudsill as wide as the footing would make no difference, nor will cutting the mudsill flush with the inside face of the cripple studs.
5. Removing the edge of a graded piece of lumber can change the grade for the remaining piece. For bending members you cannot rip two inches off of a Number 2 section of 2x6 and end up with a Number 2 piece of 2x4. The narrower piece of lumber may

have knots that are now next to its edges and would severely degrade the strength of the 2x4. However, older mudsill stock is typically clear, close-grain redwood, which is about as uniform in quality as you can find in lumber. Furthermore, mudsills are not loaded in bending, and the most axial force that the mudsill could accumulate is what builds up between anchor bolts.

6. Edge distance for new foundation anchors is a valid concern. There is no question that reducing the width of the mudsill forces a corresponding reduction in width available to place foundation anchors. This concern is best left to the installer: are they willing to use sufficient smaller-diameter anchors in exchange for cutting the mudsill flush

Appendix E
Problems with "Angle Iron" Connections

This appendix presents my professional opinions as of 2012 on the use of "angle-irons" as part of earthquake retrofits in wood-framed buildings.

Summary

A relatively common method for retrofitting wood-framed buildings to better resist earthquakes involves installing "angle-irons" that connect the existing floor framing and footings together (see Figure 1). Since the time angle-irons were brought into common use, connectors such as Simpson Strong-Tie's "UFP10" and KC Metals' "RFA136" were developed. These proprietary products are specifically designed to connect framing members to the foundation under conditions with restricted access. Angle-iron connections have never been tested, often require project-specific engineering, and require the designer to check many possible failure modes. In my experience the variety of failure modes and ineffective installations casts suspicion on all angle-iron installations. While it is possible for engineers to design angle-iron connections that would adequately resist imposed earthquake forces, conditions under old houses rarely allow installation as specified. I recommend that this sort of connection be abandoned except when specifically engineered as a "secondary anchor" for a hillside home.

Figure 1. Steel angle connecting floor framing to concrete footing.

In this paper I describe forces originating in the building and being resisted by the footings; in reality the inertial forces result from the ground moving under the building. There are no practical implications of this assumption, but I wish to acknowledge those who know the difference.

I wish to make it clear that some very respectable and diligent engineers and contractors use angle-irons. This paper is not meant to suggest in any way that these people are irresponsible or unethical. My frustration is more with the code-development process which has, in my personal opinion, left a vacuum that has been partly filled by this and other untested methods.

1. Background

In August 2011, I was retained to evaluate a retrofitted 1960's era building where angle-irons were used to connect 4x6 floor joists to the low concrete stemwalls. The angle-irons had been carefully installed; yet when I calculated the allowable loads on them I determined that many supplemental connectors were needed in order to attain a satisfactory level of strength. The angle-irons provided only about 20 percent of the strength needed to meet IEBC retrofit guidelines.

The surprisingly low capacity I calculated for the angle-iron connections increased my concerns about this method. Private home inspectors see many instances of poorly-applied or ill-conceived angle-iron connections, but it became clear that even well-executed jobs may often be inadequate.

2. Connection Description, Intent, and History

Figure 1 shows an installation where a steel angle connects a wood floor beam to the concrete footing. The typical installation is an angle section with 3-inch wide by ¼-inch thick "legs" with two bolts connected to the concrete and one or two bolts connected to the floor framing member.

This connection is intended to resist side-to-side earthquake forces as well as uplift. Each installation may have unique dimensions or conditions that require specific analysis, as opposed to stock manufactured hardware such as Simpson Strong-Tie's "UFP10" or KC Metals' "RFA138" (which can have their own installation problems—however, such problems are usually easier to notice).

I am not aware of the specific origins of the steel angle connection. Reports from various professionals suggest increased use after the 1989 Loma Prieta earthquake. At that time there were no manufactured connectors to serve the purpose of the steel angle as shown in Figure 1.

3. Practical Advantages

Under ideal installation conditions, the advantages of this connection include:

- It can be installed to shallow floor framing members where restricted access would otherwise prevent connections to (or through) the existing mudsill
- It could be effective in resisting forces in all three directions (parallel or perpendicular to the footing, and upward)
- Material costs are low (this may be offset by fabrication costs).

4. Field Installation Problems

Some framing conditions may be conducive to using angle-irons as a retrofit method. However, variations in field conditions make effective use difficult to achieve. There are many problems that might arise, any of which could severely impact the connection's effectiveness. A few of these are listed below.

4.1 Misunderstanding of the nature of earthquake forces. Many of the installations shown here may have occurred because the installer did not know that horizontal forces cause more failures than uplift (hardware of all sorts is misused to resist uplift rather than lateral forces; the following demonstrate this misunderstanding as applied only to angle-irons).

4.2 Creative mis-installations. The demand for earthquake retrofits in the wake of Loma Prieta and Northridge led to contractors developing their own retrofit methods or copying methods that they had seen in other installations. Even minor changes to installation methods can greatly reduce the effectiveness of the system. Figure 2 includes just a few examples of cases where installers went to great effort to install hardware whose purpose was not well understood.

Figure 2A. Connection to new blocks where existing joists run parallel to footing leads to a very indirect load path.
Photo: Paul Rude, GGASHI

Figure 2B. Wood spacer between footing and steel angle severely reduces lateral strength.
Photo: Paul Barraza, GGASHI

Figure 2C. A single bolt to the footing provides no lateral capacity at all.
Photo: Paul Rude, GGASHI

4.3 Ignoring leverage: The total height of the angle with respect to the distance between the two bolts that connect it to the footing is critical. Leverage on the bolts connecting the angle to the footing is often underestimated, ignored, or completely misunderstood in cases where the installer or designer intended only to resist uplift. Even the most favorable installation conditions usually double the lateral force that the top footing bolt must resist in relation to the earthquake force imposed at the top of the angle-iron. Figure 3 shows several examples of angles that have little hope of preventing movement before shearing off the top anchor into the foundation.

Figure 3A. This angle has about a 5-to-1 lever ratio on the top foundation bolt.
Photo: Paul Rude

Figure 3B. Leverage is combined with a gap between the angles and concrete where the electrical cable was not moved. Photo: Paul Rude

Figure 3C. The tallest angle installation I have seen to date.
Photo: Paul Barraza

4.4 Angle size. Angles are rather inefficient bending members. Bending failure can occur at surprisingly small loads if angle legs are not at least 3" wide by ¼" thick, or if the angles exceed moderate length—say about two feet total.

4.5 Gaps between angle and concrete: The footing surface must allow the steel angle to be in direct contact with the footing at the point where the bolts attach the angle to the concrete. Gaps at the connection points lead to two main problems. First, it is very easy to overstress the bolt; bolts loaded sideways lose significant capacity if the load is not applied adjacent to the concrete surface where the bolt is embedded. Second, bending of the bolts in the concrete allows the steel angle to lean. Even slight movement at the connections to the concrete is magnified at the floor level based on the angle height versus the distance between the footing bolts. Figure 4 shows three cases of gaps at footing bolts; Figure 2B shows another case.

Figure 4A. Approximately 3/8" gap between steel and concrete at bolted connection.
Photo: Thor Matteson

Figure 4B. Armored electrical cable; one of many possible obstacles that could cause a gap.
Photo: Paul Rude, GGASHI

Figure 4C. Wood spacers create a 3-inch gap. Specific reason unknown, but many obstacles await retrofit workers.
Photo: Paul Barraza, GGASHI

Because so many factors can affect the forces on bolts and the impact that deflection could have on the system's capacity, each case requires specific analysis; even small gaps should be investigated before the connection is assumed adequate. Such analysis is very time-consuming, and often it makes more sense to just install a new proprietary connector to resist lateral forces.

4.6 Placement of steel angles can occur only at joists. This allows for less adjustment than hardware such as the UFP10, which can be located at any convenient point along the mudsill to avoid obstacles or uneven concrete surfaces (the RFA138 connects to the top of the mudsill, thus being less versatile). At best the angle-iron can be switched from one side of the joist to the other, but this requires drilling additional bolt holes to allow for left- or right-handed installation.

4.7 Steel angles are not readily adjusted in the field. Many of the installation problems shown here might have been avoided if new holes could have been drilled through the angle-iron, or the angle-irons had been shaped to fit the footing. Retrofit installations may need anywhere from 10 to 60 angle connections; measuring for each case is simply not practical. Yet prefabrication leads to ineffective installations because no adjustments can be made to meet circumstances found in the field.

5. Engineering Advantages

Distinct advantages of steel angles over other currently-available retrofit hardware and methods are not evident to me. Using the same connection to resist forces in both the parallel-to-footing and perpendicular-to-footing directions could be appealing (my disagreement with this assumption is explained later). Ability to resist uplift is not needed for installations without cripple walls, and the connection is not made in the location needed to address overturning in walls above the framed floor level. I have seen so many bad variations of angle-iron connections that I feel the disadvantages far outweigh the advantages.

6. Engineering Concerns

I was recently retained to evaluate a prior retrofit that used steel angles (examples shown in Figures 1, 4A, and 5). I was lucky in this case that every steel angle had identical dimensions and bolt placement. The installation was executed very well, but I ended up giving the steel angles very little credit toward resisting lateral forces. The following concerns come to mind; the first four are strictly from the viewpoint of evaluating existing installations, the others are structural problems that apply to the evaluation of existing installations and design of new installations.

6.1 Each installation must be analyzed individually. (It is unlikely that I will ever come across a case identical to the one I recently evaluated.) It is very possible that several configurations must be evaluated for a single building.

6.2 Even identical pieces of steel angle require separate evaluation if they are placed differently on the joists, have gaps between them and the footing, etc.

6.3 Deflection is magnified by connection geometry. Assuming that the angle-iron pivots around the bottom footing bolt, then any slippage at the top footing bolt increases in proportion to how far the angle-iron extends above the top bolt. Angle-irons are often installed with the distance between the two footing anchors being equal to distance from the joist connection to the top footing anchor (see Figure 6). Standard steel fabrication practice is to size bolt-holes 1/16-inch larger than the bolt diameter. Together with the preceding, movement at the floor framing could reach 1/8-inch due to bolt slippage alone. Contrast this to Simpson's testing of the UFP10, where their listed capacity is based on the load when deflection reached 1/8-inch. Using this same criterion for establishing load capacity of angle-irons would give essentially zero.

6.4 Relative stiffness of the installation is affected by many factors (ratio of bolt separation, possible cantilever of bolts from footing, oversized holes in the steel, etc.) Figure 3C shows an installation that is already pitifully inadequate; even if the connections and materials were able to resist the imposed forces, the angle-irons would not act together because of their different stiffnesses.

6.5 No testing has been done to verify the calculated capacity of the system. Given the many variables that affect performance, the predicted capacity could vary greatly from test results.

There are several failure modes that must be checked for each installation. Some of the most obvious include bending failure of the angle, failure of the anchors to the footing, bearing capacity of the washer against the side of the joist, and cross-grain bending in the joist itself. The most insidious of these failure modes is cross-grain bending of the joists, which is discussed at greatest length here.

6.6   Cross-grain bending. Lateral forces resisted by the steel angles transfer from the bearing area of the angle or plate washer against the face of the joist to the floor diaphragm attached to the top of the joist. For loads acting perpendicular to the joist direction, the eccentricity between the top of the joist and the top of the bearing point induces cross-grain bending in the joist (see the attached two pages of calculations).

Cross-grain bending is not mentioned in the building codes except to say that it "should be avoided" if possible. The National Design Specification for Wood does not list an allowable stress value for cross-grain bending, with the intent of further discouraging reliance upon it. Correspondence with the American Wood Council suggests the maximum "allowable" stress for cross-grain bending would be one-sixth of the current allowable shear stress (currently 190 psi for Douglas fir).

[As a side note, stock hardware is sometimes used in ways that would induce cross-grain bending. The most common instance of this is where Simpson "H10" or similar ties are used to transfer diaphragm shear forces to the top plates of cripple walls, or potentially directly to mudsills.]

In cases where the joists are held in place by continuous, snug-fitting solid blocking, it is conceivable that the lateral load path would be through the blocking into the joists, with the joists acting in weak-axis bending between the blocks and the angle-irons to transfer loads at that part of the load path. In cases where blocking is not present (or not continuous), or a rim joist is present instead of joist blocking, cross-grain bending failure is likely to be the governing failure mode in an angle-iron connection.

The NDS clearly advises against any reliance on cross-grain bending. Individual designers should be entrusted to provide designs that do not induce cross-grain bending. Such designs would likely rely on existing joist blocking to prevent joists from twisting. Having seen the enormous variability in construction under even a single house,
I personally would not expect solid blocking to be present in all instances even if it was verified in certain under-floor areas.

6.7   Assumption of force distribution in diaphragm. As stated earlier, some installations assume that angle-irons will resist earthquake forces in both directions, i.e., perpendicular to the joists as well as along their length. Figure 5 shows a case where no retrofit connections are placed along footings parallel to the floor joists.

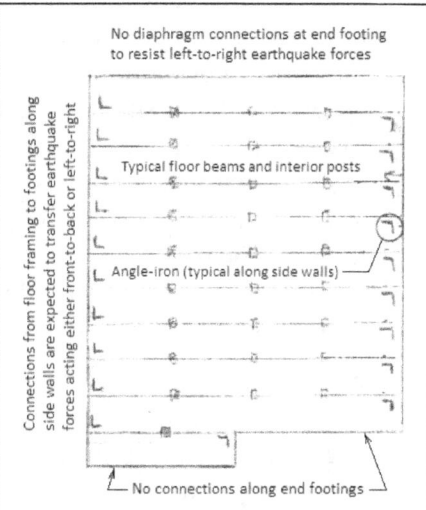

Figure 5. This schematic floor plan shows a design that relies on the steel angles along the side walls to resist earthquake forces acting in any direction.

Earthquake forces in walls above the first floor diaphragm do not have a direct load path to the foundation: the loads must transfer into the floor diaphragm, spread out uniformly, and find their way to the angle-iron connections to the footing.

Most older homes are divided into many rooms and have multiple cross-walls in both directions. The many walls could distribute loads from the roof and second floor levels above to the first-floor diaphragm fairly well under the right circumstances. However, depending on the layout of the building above the first floor, it could be optimistic to assume that earthquake loads will follow such an indirect load path. In addition, the relative stiffness of interior versus exterior wall finishes (gypsum board or plaster vs. stucco, etc.) could significantly affect the way earthquake loads travel through the building. Rather than hope that lateral forces in end-walls parallel to the floor framing will find their way to angle-irons spread along the side-walls, it seems most prudent to provide a direct load path from the end walls directly to the footings beneath them.

Figure 6 shows a variation on the straight angle-iron that is better suited to low-clearance installations, or where there is not sufficient stemwall height for angle attachment. A rectangular steel plate, about 16 inches wide and 8 inches high, is welded to the back of the angle; the plate is then bolted to the footing. This allows a pair of bolts to be installed closer to the top of the footing, reducing the need to excavate out for a longer section of angle. However, it totally negates the ability of the connection to resist forces perpendicular to the foundation. The plate could easily bend under load, and the bolts (especially when placed as shown) are subjected to significant prying actions when a force acting away from the footing is applied at the top of the angle.

6.8   Deflection of the system.   Even if the joists that the angles connect to are blocked to prevent them from rolling over and succumbing to cross-grain bending, then deflection of the floor framing system with respect to the footing will induce prying action between the steel angle and the joist. With oscillation back and forth, the connection would become increasingly loose. The effect of this loss of capacity with repeated cycles would be anything but good.

Figure 6. This modified angle-iron connection eliminates the potential for resisting forces in the direction of the joist because prying action and plate bending are not accounted for.
Photo: Paul Barraza, GGASHI

7. Conclusions

The best quality angle-iron retrofit I have seen so far is the one I was retained to evaluate. The house is single-story with stucco exterior, built in the early 1960s with 4x6 floor beams 4 feet apart. Steel angles had been installed on one side of every floor beam. After checking the several possible failure modes, I had to recommend installing almost as many UFP10s as would have been needed for a completely unretrofitted house. My client paid about a thousand dollars for me to investigate the existing work under the house and then analyze its effectiveness. This fee represented about 25 to 35 percent of the total spent on retrofitting a house that had already been retrofitted.

In the future I will suggest that clients completely disregard any existing angle-irons; their money would be better spent on installing new hardware that is specifically designed and manufactured for the purpose, with test results to support the rated load values. An exception to this recommendation would be in cases where the installation followed an engineered design and the design included analysis of all the failure modes listed above or in the attached calculations.

8. Recommendations

This section contains two sub-categories: guidelines on how to quickly evaluate existing angle-iron installations, and how policy makers and building officials might guide their future use to avoid installations like those shown or described in this paper.

8.1 Evaluating existing installations

This section guides non-engineers in determining whether existing angle-irons are completely useless (some engineers would say downright harmful), or if they may provide some level of protection against earthquake damage.

Defective installations:

If any of the following apply, an individual steel angle connection is essentially worthless:

1. Only one bolt installed from the steel to the footing.

2. Wood spacers or any other shims or gaps more than 1/16" (the thickness of a nickel) between the steel and the footing.

1. Distance from the top bolt in the concrete footing to the bolt in the floor joist is greater than the distance between the two bolts in the concrete (See Figure 7).

2. Joist blocking is not present on both sides of the joist that the angle connects to; blocking must be tight-fitting and the same depth as the joist. [Floor framing in "budget" construction was often installed with blocks in every other joist bay; in that case new blocking would be needed. If a rim joist was installed, blocking is often not present at all.]

3. Either of the 'legs' of the angle is less than 3" wide, or less than ¼" thick.

4. Anchors connecting the angle to the concrete are less than ½" diameter.

5. Standard round washers are used under the bolt head or nut against the wood where the angle connects to the joist, or no washer is present at all.

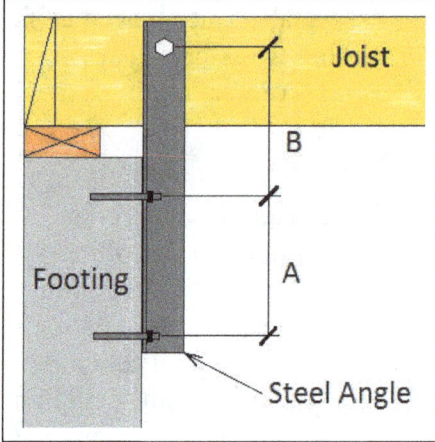

Figure 7. Relative placement of bolts is critical to performance. For passable performance, dimension "A" must be at least the same as dimension "B."

Even if individual steel angle connections pass all of the above requirements, the overall installation may be inadequate in either of the following cases:

1. Configuration of steel angle connections is not consistent throughout the crawlspace, i.e. the distance between footing bolts varies from one angle to another.
2. If the house is more than a single story, or if it has large rooms at the ends, retrofit connections should be installed along all mudsill and framing connections—not just the ones where joists are perpendicular to the footings.

Requirements for acceptable installations: In rare cases steel angle installations may provide enough strength to resist earthquakes. In my opinion, any such installation would require individual analysis by an engineer that includes all of the potential failure modes as addressed above plus any others that may affect the particular installation.

Cross-grain bending in the joists is probably the most overlooked failure mode. Joist thickness has a huge effect on cross-grain bending stresses. Cases below are divided into 4x joists/beams and 2x joists.

*Connections to **4x** floor framing **may** be adequate under the following conditions:*

1. Joists sandwiched between pairs of steel angles when the angles are installed with their tops within ½-inch of the underside of the floor sheathing
OR:
Steel angle and minimum 2" square, 3/16-inch-thick plate washer at angle-to-joist connection are both installed within ½-inch of the underside of the floor sheathing.
2. Maximum spacing of steel angles is 32" for single story houses, 16" for two-story houses.

*Connections to **2x** floor framing **may** be adequate under the following conditions:*

1. Joists sandwiched between pairs of steel angles that are installed with their tops within
¼-inch of the underside of the floor sheathing
OR:
Steel angle and minimum 2" square plate washer at angle-to-joist connection are both installed within ¼-inch of the underside of the floor sheathing.
2. Maximum spacing of steel angles is 32" for single story houses, 16" for two-story houses.

8.2 *Requirements for new installations, if allowed at all:*

Since so many things can go wrong that greatly impact the connection's strength, I suggest that angle-iron installations be subject to special inspection or structural observation to verify at least the following:

1. All dimensions of system correspond to those used in design, including but not limited to: distances between bolts, locations of bolts in joists, edge-distances in footings or joists, gap between angle and floor sheathing above, thickness of joists, end distances in joists, steel angle size, plate washer size.
2. Snug-fitting, continuous blocking present on both sides of all joists with steel angle connections.
3. Condition of existing joists to verify wood is sound.

4. Substantial contact between steel and concrete at steel angle connection points to footing.
5. Steel angles in firm contact with face of joist.
6. Plate washers used against wood under nut or bolt.

ANGLE-IRON CAPACITY

L3×3×1/4×1'-8"   S = .569
5/8" ⌀ BOLTS

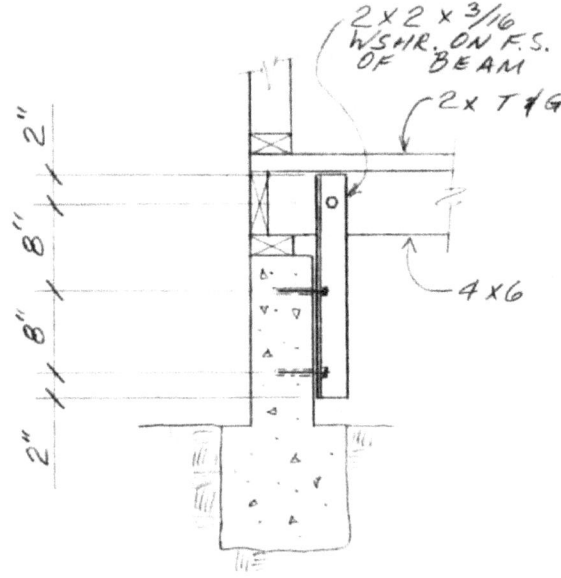

CHECK FOLLOWING FAILURE MODES FOR LOADING IN LINE W/ FOOTING
1. STEEL ANGLE BENDING
2. BOLTS IN FOOTING
3. WOOD BEARING AGAINST SQ. WSHR
4. CROSS-GRAIN BENDING IN JOIST (EXPECTED TO GOVERN)

1. $M_{MAX} = F_b S = 36(.6)1.33(.569) = 16.4$"k
   $M = P(8")$ $P_{MAX} = 16.4"k / 8" = 2,050\#$

NOTE: THIS CONNECTION IS OFTEN SEEN WITH ANGLE STOCK AS SMALL AS L2×2×3/16, S=.188; LONGER LEVER-ARMS ARE ALSO COMMON.
FOR L2×2×3/16, $P_{MAX} = 677\#$

2. BOLTS IN CONCRETE: FOR GEOMETRY SHOWN, CAPACITY IS 1/2 OF THE CAPACITY OF THE TOP BOLT.

2. (CONT'D)
FOR HIGH-QUALITY CONC.
5/8" HILTI-HY CAP. = 2460#
CAP. = $\frac{2460}{2}(1.33) = 1,640\#$

3. WOOD BEARING:
$P_{MAX} = 625 psi [(2in)^2 - .78(\frac{11}{16})^2]$
$P_{MAX} = 2270\#$
(FOR ROUND WASHER: $.78(.625)(1^2 - \frac{11}{16}^2)$
$\approx 260\#$)
CHECK WASHER BENDING:
$M = .625(2)(0.5) = .625$ in-k
$S = (2 - \frac{11}{16})(\frac{3}{16})^2 / 6 = .008$

$M_{ALLOW} = .008(.8)36 = .221$ in-k

$P_{MIN} = 2,270 \left(\frac{.221}{.625}\right) = 804$
MORE REASONABLE TO SAY
$P_{ALLOWABLE} = \frac{1}{2} 2270 = 1,130$
SINCE 2-DIMENSIONAL BENDING OF PLATE-WASHER UNDER NUT

4. CROSS-GRAIN BENDING:
FOR BOLT @ MID-HT. OF JST. AND CONSIDERING JOIST SECTION OF 1.5 TIMES THE BOLT END-DISTANCE:
$M_{MAX} = F_B S$
$S = \frac{1.5(6")3.5^2}{6} = 18 in^3$
$F_{B \; CROSS-GRAIN} = \frac{1}{3} F_v = \frac{1}{3} 95 psi (1.33)$
$F_{BCG} = 42 psi$
$M_{MAX} = 18 in^3 (42\#/in^2) = 760$"#
P_{MAX} @ 2.75" = $\frac{760}{2.75} = 280\#$

COULD BE GENEROUS AND USE INCREASED F_v PER RECENT NDS, OR M BASED ON DISTANCE FROM TOP OF JOIST TO TOP OF SQ. WSHR OR "CLAMPING" DUE TO WALL D.L.

SEE NEXT PAGE FOR MORE DETAILED ANALYSIS

CROSS-GRAIN BENDING:

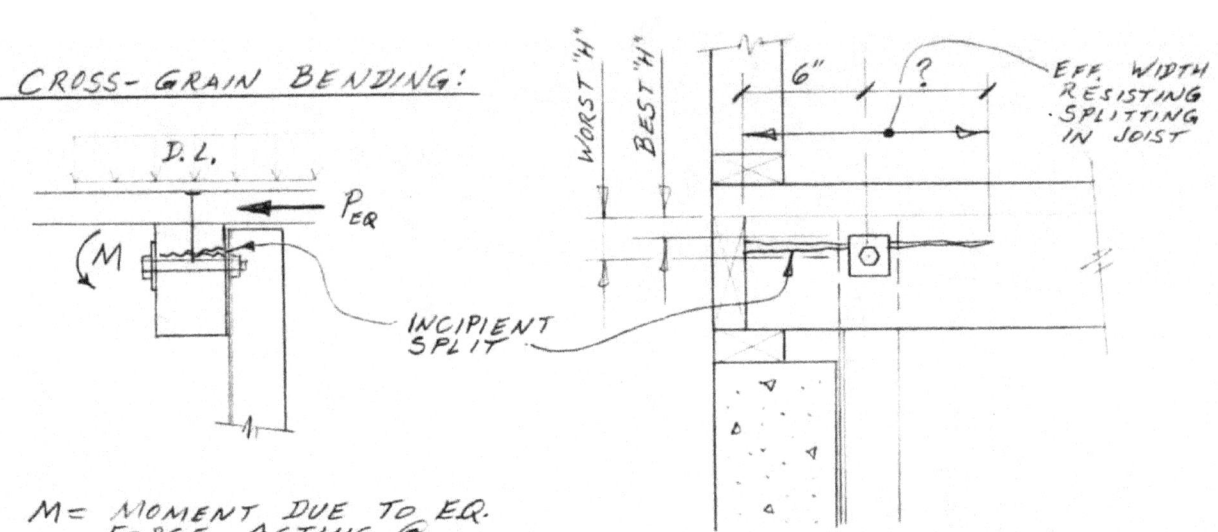

INCIPIENT SPLIT

EFF. WIDTH RESISTING SPLITTING IN JOIST

M = MOMENT DUE TO EQ. FORCE ACTING @ ¢ TOP OF BEAM $ RESISTED BY BOLT/WSHR. AT "H" INCHES BELOW TOP OF JOIST

H — VARIES; BEST CASE FOR TOP OF BOLT @ 2" FROM TOP OF ANGLE & JOIST, USING 2" SQ. WASHER THAT IS ASSUMED TO PROVIDE RIGIDITY AT TOP OF WASHER: $H = 1"$

LESS OPTIMISTIC: $H = 1\frac{1}{2}"$

MORE REALISTIC, SINCE ANGLE NOT ALWAYS TIGHT TO UNDERSIDE OF FLOOR: $H = 2"$ TO $2\frac{3}{4}"$

"CLAMPING" ACTION DUE TO WALL WT:
FOR 8' STUCCO WALLS, 8' TRIB ROOF:
DL = 8'(18 PSF + 10 PSF) = 224#/FT OF WALL
FOR JSTS. @ 16", 300#/JST. — 900# @ 4'
— SAY ½ BEARS ON RIM JST.

DISTRIBUTE OVER "SPLIT" AREA:
450#/(3.5)(9) = 14 PSI

BEST-CASE SCENARIO:
$F_V = 190(1.33) = 253$ psi
$F_{B_{CROSS-GRAIN}} = \frac{1}{3} F_V + 0.9(14 \text{ psi})$
$F_{BCG} = 97$ psi

FOR DETERMINING "S" EFFECTIVE IN RESISTING CROSS-GRAIN BENDING:
$$S_{EFF} = \frac{bd^2}{6} = \frac{9(3.5^2)}{6} = 18 \text{ in}^3$$
WHERE: d = JST. THICKNESS
b = EFFECTIVE WIDTH = 6" + ? (9" USED)

BEST-CASE (CONT'D.)
$M_{MAX} = S_{EFF} F_{BCG}$
$= 18 \text{ in}^3 (97 \text{ psi})$
$= 1,740$ in·lbs

$P_{MAX} = \frac{M}{H} = \underline{1,740\#}$

FOR $1\frac{3}{4}"$ THK. JOISTS,
$H = 2", F_V = 95$ psi:
$P = \frac{9(1.75)^2}{6} \left(\frac{95}{3} + \frac{300}{1.75(9)} \right) / 2"$

$P = 4.6 \text{ in}^3 (51 \text{ psi})/2"$

$P = 115\#$

HUGE VARIABILITY BECAUSE SO MANY FACTORS AFFECT RESULT.
PER CODE: <u>CROSS-GRAIN TENSION SHOULD BE AVOIDED WHEN POSSIBLE</u>

Appendix F
Severe Roof Weakness Present in some pre-1920 Houses

Most roofs overhang the walls of the house. An interior view of the most common framing method for constructing the overhanging portion of the roof is shown in Figure 1 and Figure 2. The rafter is cut to fit onto the top plate of the wall, and extends beyond the support to create the eave overhang. Ceiling joists span across to the rafters on the opposite side of the house and tie pairs of rafters together so they do not collapse and push the walls apart. Some sort of tie across the house is essential to keep this type of roof standing. A detailed discussion of roof framing is outside the scope of this book beyond saying that the rafter ties (usually joists) should be as close to the bottom of the rafters as possible. "Collar ties" near the peak of the roof may be needed to resist the roof opening at its peak due to wind uplift, but they do not resist the outward thrust at the base of the rafters.

Figure 1 GOOD: Typical roof rafter connection to wall and ceiling joist used in contemporary construction. The ceiling joist ties rafters on opposite sides of the house together so the walls do not spread apart. The "face-nailed" connection from the joist to the rafter is the most secure type of nailed connection.

Figure 2 Typical rafter and ceiling joist connection with no soffit at eave.

The weakness under consideration occurs when builders wanted a level "soffit" (underside of the roof overhang). They extended the ceiling joists out from the walls to create an overhang, as shown in Figure 3. They then supported the rafters on the ends of the extended joists.

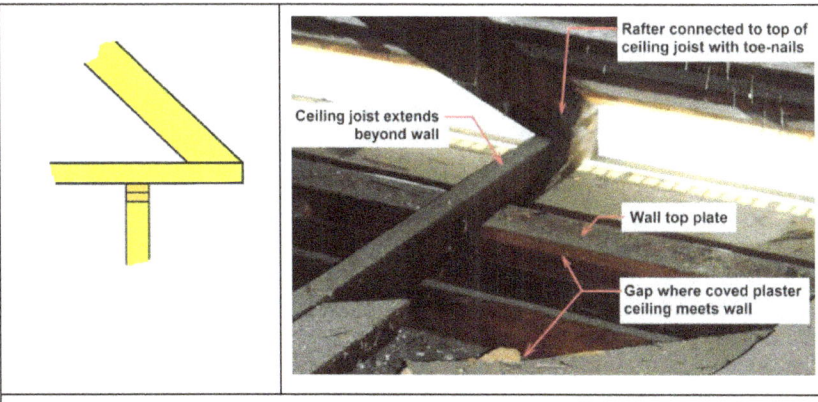

Figure 3 BAD: Archaic roof connection using extended ceiling joists to create soffit at eaves. The rafter connection to ceiling joist is marginal. Photo: Max Curtis, GGASHI

This method introduces a few problems, two of which apply to earthquake resistance. The primary concern is the connection between the rafters and the joists that tie them together. In the current method, the rafters and joists lap against each other and are connected with "face nails" between the members. In the archaic method, the connection was made with "toe-nails" driven from the rafters down into the joists. Toe-nails do not provide nearly the connection capacity of face nails. If this connection fails, the roof or portions of it could collapse onto the ceiling framing. This would likely result in significant damage to the house interior. In any case your house would no longer be protected from weather.

Many old houses have several layers of roofing on them, which can add several tons of material to the roof. The roof is the worst place to add weight to your house—especially if you have antiquated connections.

The obsolete framing method also lacks a direct load path from the roof diaphragm into the side walls of the house. I consider this less hazardous, but still a good thing to fix. One effective repair method would be installing "shear frames" between sets of rafters and joists. This method is described in Section 3.2.2 of "Wood-Framed Shear Wall Construction—an Illustrated Guide".

How to identify the condition from the exterior

This particular weakness is created when a level soffit overhangs the walls. However, a level soffit does not always indicate the weak connection. Figure 4 and Figure 5 show a secure rafter connection with a level soffit. The key difference is the soffit is dropped below the adjacent ceiling height.

Figure 6 shows an early 20th century house with a level soffit at the ceiling height. Compare the distance from the soffit to the tops of the windows for the houses in Figure 5 and Figure 6.

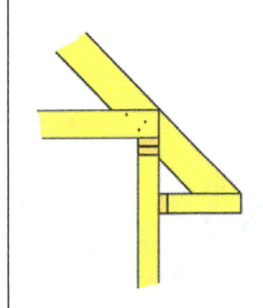

Figure 4 Framing method for "dropped" level soffit.

Figure 5 Dropped soffit (note level just above tops of door and window).

Figure 6 Level soffit at ceiling height. Note wide band between soffit and tops of windows. Photo: John Fryer, GGASHI

Appendix G
Excerpts from *Assessment of Damage to Residential Buildings Caused by the Northridge Earthquake**

*A report released by the U.S. Department of Housing and Urban Development, Office of Policy Development and Research, July 30, 1994, available online here: http://www.huduser.org/portal//Publications/PDF/earthqk.pdf

The above document gives a good summary of findings related to the Northridge Earthquake.

One important comment relates to roof damage.

"The photograph below shows an extremely rare case of a collapsed roof. Roof structural damage was almost nonexistent in the single-family dwelling housing population, with the exception of localized roof damage caused by masonry chimneys. Perhaps the large vertical component of ground acceleration contributed to this roof's failure. More likely, it was the lack of properly sized and spaced rafter ties for the added dead load of a tile roof covering, as required by the LA City "Type V" prescriptive requirements. When found, roof structural damages were most often precipitated by alterations or renovations to the homes which modified, removed, or overloaded the rafter ties."

"Extremely rare" case of roof collapse, likely due to heavy tile roofing installed after the original construction, possibly in conjunction with removal of original structural members.

[Author's note: "Type V" construction refers to construction used in almost all single-family homes. For decades, building codes have divided construction into five "types" using Roman numerals I through V. The construction types have specific requirements for building materials, structural systems, property-line setbacks, fire protection, maximum floor area, number of stories allowed, etc. Type V is the least restrictive of these. In this context, "Type V" simply means "wood-framed"]

CONCLUSIONS AND RECOMMENDATIONS

Results of this study support the following conclusions:

- Although new homes are commonly engineered in the Greater Los Angeles region, the majority of existing single-family homes in the San Fernando Valley area were built to prescriptive methods known locally as "Type V" *[see author's note above]*. This manner of construction is also common to older multifamily buildings.
- Structural damage to single-family dwellings was infrequent, and performance was generally very good.
- Structural damage was primarily located in the foundation systems where less than two percent of the single-family dwellings suffered moderate to high levels of damage to their foundations.
- Most occurrences of moderate to high foundation damage were associated with localized site conditions including liquifaction, fissuring, and hillside slope failures.
- Damage to wall and roof framing in single-family dwellings was limited to low damage on about two percent of the walls and less than one percent of roofs.
- Case studies of extreme damage to single-family dwellings reveal that severe structural damage to foundations, walls, and roofs, existed, but at extremely low levels of occurrence.
- Finishes experienced the most widespread damage, with 50 percent of all single-family dwellings suffering at least minor damage. However, only 4 percent or less could be classified as moderate to high damage. Most finish damage was related to stucco and drywall/plaster cracks at the foundation or at openings in walls.
- Statistical inferences were generally inconclusive. One exception was a significant difference in damage levels to exterior finishes on single-family dwelling single-story homes with crawlspaces versus slab-on-grade foundations. Homes on slab foundations suffered some degree of damage to exterior finishes in about 30 percent of the sample, while crawlspace homes approached a 60 percent damage rate.
- Although not captured in the random statistical survey results, it was observed in the field that:
- Case studies of multi-family low-rise and single-family attached homes indicate that soft-story garage construction is particularly susceptible to severe damage. In a few cases, severe racking led to total collapse of the soft story. When properly detailed and constructed, newer buildings with stronger foundations and plywood wall sheathing seemed to perform well.
- As with single-family dwellings, cracking of stucco and interior finishes on single-family attached and multi-family low-rise dwellings was widespread, particularly at openings in walls.

- Damage to masonry chimneys and fireplaces was common but not consistent. In many cases, movement of masonry chimneys caused localized damage to other parts of the building. Prefabricated wood chimneys with metal flues did not appear to be similarly affected.
- Damage to masonry privacy walls appeared to be widespread. Many of these were unreinforced walls.
- Damage to contents appeared widespread in all types of homes.

The following recommendations are offered based on results of this study:

- Based on the varying performance of different types of construction, "across the board" code changes should be avoided. For example, it may be necessary to increase the racking resistance of larger attached buildings, but results indicate that single-family dwellings built to the older, less stringent prescriptive ("Type V") requirements performed well.
- Performance requirements or prescriptive requirements in building codes should consider solutions addressing the unique seismic risks associated with construction on sloped sites (hillsides) in active seismic regions.
- Prescriptive building code structural requirements should be expanded to include cost-effective solutions addressing the added seismic risks associated with the architectural features common to newer homes (e.g., cathedral ceilings and large wall openings).
- Inspections and retrofit should be considered for older soft-story construction, especially where living spaces are located in the soft story.
- Future studies should attempt to assess the amount of damage to contents and nonstructural elements relative to overall building structural damage.
- The methodology and data sheets used in this study and the previous HUD-sponsored study of Hurricane Andrew and Iniki should be examined to identify improvements in the sampling and analysis methods.
- Research should be conducted into alternative garage foundation construction methods for multi-family low-rise buildings which will absorb greater amounts of energy in a non-destructive manner. For example, heavy, treated timber construction may be a viable alternative which would be compatible with wood-framed upper stories and also reduce lateral loads on the upper stories.
- Results of this report should be widely disseminated to policymakers and building authorities of communities located in the active seismic regions of the U.S.
- Statistically-valid damage estimates presented in this report should be used to help determine costs and benefits related to building code modifications, policy decisions, and other actions that affect the seismic hazard to homes.

Appendix H
Excerpts from *Practical Lessons from the Loma Prieta Earthquake* (National Research Council)

Washington DC: National Academy Press, 1994

Selected commentary regarding apportioning of research funding and the disconnect between what engineers expected versus what the public expected (emphasis added)

FOCUS OF RESEARCH

Disasters tend to generate a frenzy of activity in the research community. There is always a renewed interest in solving the problem and generally a limited amount of resources to fund the needed research. After the Loma Prieta earthquake, the August 1990 Earthquake Engineering Research Institute newsletter reported that the National Science Foundation and the U.S. Geological Survey awarded approximately $4.1 million in grants to do various studies. While learning from earthquakes, researchers should be studying the effects that are unique to that earthquake, yet it is interesting to note the focus of those grants. Twenty-four percent were focused on soil related topics, with an additional 10 percent on site-response issues. In sum, about 35 percent focused on geological and seismicity issues. **Five and one half percent addressed evaluation and retrofit issues**, and 2 percent addressed unreinforced masonry. Of note is that 2 percent of the awards money went to housing risk observations. The largest of the grants, 3.3 percent of the money, went to the University of Colorado for work on "The purchase of Earthquake Insurance in California." Finally, 78 percent of the grants went to individuals associated with a university, and 10 percent went to practicing design professionals.

This summary of research awards is not a critique but is presented to focus the attention of the reader on evaluating what research seems to be needed as a result of the earthquake and where and how that effort might be best spent.

For research to be effective, it is essential that it be focused on the goals of a program. If one views research as the advancement of knowledge, then the studying of a subject for better understanding will in time be a benefit to the analysis and design of buildings. However, designers often have little patience with the research efforts, because the results do not translate directly into code language. There are short-term demands for research, such as the California Department of Transportation (CALTRANS) retrofit studies, which must be accomplished on a priority basis. Beyond those types of studies, **there needs to be a more direct effort with a nationwide focus that will lead to advancements in knowledge in areas where it will have the most impact**.... *[Author's note: this is still true 21 years later for wood-framed buildings]*

...

- *Researchers* seek analytical solutions or definitions of parameters from a postulated problem.
- *Engineers* seek a better definition of performance of materials.
- *Building officials* seek a prescriptive document, which is primarily black and white.
- ***Building owners* wonder what they are paying for and expect a functioning building in the post earthquake environment.**

...

PRINCIPAL FINDINGS

Seismic Risk

Sixty-three people are dead, and engineers are saying that structures performed about as expected. In excess of 27,000 buildings are damaged, and engineers are saying that is about what was expected. Public schools survived virtually unscathed, and again engineers say that is about what they expected. This was the first real confrontation with the performance intent of the modern seismic code. Before the Loma Prieta earthquake, "earthquake proof" was still a term that was used. Post-earthquake things have changed. **The realization that buildings could be damaged and still be in complete compliance with the building code was now at hand.** What were the expectations of the general public and the building owners?

Regardless of the expectations, there was a new reality. Buildings were not designed to be functional after earthquakes, they were designed only to be life safe. Building owners weighed the consequences of that, and soon the financial and insurance industries also saw the broad implications of such performance. It also became clear that retrofits for less than code levels were targeted for life safety only and that preservation of a historic resource was not the intent of applying the less than code approach. Retrofitting of historic buildings for less than *preservation performance* risks the loss of the assets...

Residential Construction

The residential problem was identified as a series of definable issues. The legislature mandated that the Seismic Safety Commission develop the *Homeowner's Guidebook to Earthquake Safety* (California SSC, 1992), which is basically a disclosure document that must be handed to the buyer of a pre-1960 house prior to sale. The document does a good job of identifying, in lay person's terms, the various hazards often found in homes. **It is disappointing that the mitigation of those identifiable hazards is not mandatory prior to the sale.**

The hazards to be mitigated for residential structures are relatively straightforward. The consequences of not mitigating the simple hazards were pointed out in post-earthquake testimony. In testimony before the Seismic Safety Commission, a fire marshal was commenting on the need to brace water heaters and the potential for disaster if this is not accomplished. In the Marina District, gas fires were posing a serious threat. The fire marshal pointed out that it was lucky the gas problem was handled quickly before a fire storm situation developed and spread out of control. This was a dramatic testimony emphasizing the need for a simple mitigation measure.

Given that over 27,000 residential structures were damaged during the Loma Prieta earthquake, one would extract the logic that residential structures and the mitigation of the observed earthquake residential hazard would be high on the investment priority list of funding agencies. **Two percent of the National Science Foundation (NSF) funding went to residential research**, although the previously mentioned top award was on the topic of earthquake insurance in California. Only the California legislature took action in passing legislation to mandate the preparation and distribution of the homeowner's guide. **It would seem obvious that the residential hazard is within reach of direct mitigation, and with some financial stimulus from governmental agencies and insurance interests, it is possible that this hazard could be wiped out.**

Appendix I
References and Further Reading

The following references were used in developing this book or may have related material of interest to readers. In general, most of the following references related to earthquakes show lots of photos of damage buildings, describe general retrofit strategies, and provide a couple of construction details. (If there was already a reference that went into as much detail as this book does, I would not have spent four years compiling the material and photos included here.)

Retrofit guidelines almost never show obstructions to installing hardware or bracing panels. The crawlspaces shown never have any pipes, ducts, gas meters or the like in the way. The following brief descriptions are intended to help serious readers decide which references may be useful for further study.

1. **Second Edition of *"Wood-Framed Shear Wall Construction—an Illustrated Guide"*** by Thor Matteson, SE (referred to as the "Shear Wall Guide"); International Code Council; distribution of printed copies now handled through Builder's Booksource in Berkeley, CA.
   - Originally written to provide detailed coverage of wood-framed shear wall construction, with contractors and builders as the intended audience. The second edition of the Shear Wall Guide is 180 pages long, with over 150 illustrations and photos.
   - Addresses many field conditions and obstacles encountered in shear wall construction. It also includes information on collectors, openings through shear walls, wood shrinkage, and other topics related to new or retrofit construction.
   - This retrofit guide is based on information in the Shear Wall Guide. The Shear Wall Guide is very helpful for understanding shear wall behavior and construction.
2. *"Bracing for the Big One"*—State of Utah, Department of Heritage and Arts; online reference: http://history.utah.gov/historic_buildings/information_and_research/bracing_for_the_big_one.html
   - Offers general advice, but unfortunately features some outdated recommendations and obsolete practices and hardware.
3. *Homeowner's Guide: Earthquake Safeguards*   APA Form Number R240A, 1997
   - This publication comes from the pre-eminent authority on plywood, but it is considerably out of date.
4. **FEMA 273** *"Guidelines for the Seismic Rehabilitation of Buildings"*—1997
   - Early recommendations that served as the basis for many local codes and guidelines. This referances covers almost all kinds of buildings, from concrete warehouses to steel-

framed skyscrapers and brick buildings; only one chapter addresses wood-framed buildings.

5. **FEMA 356** *"Prestandard and Commentary for the Seismic Rehabilitation of Buildings"*—2000
    - Essentially the same as FEMA 273, but with commentary included.
6. **FEMA P-593** *Seismic Rehabilitation Training for One- and Two-Family Wood-Frame Dwellings*—2008
    - Excellent expansion of topic into specific construction methods
    - Very useful color-coded chart showing common hazards to look for; chart gives references of where to find more info (many of which are listed here)
    - Needs to be expanded from current format, which is essentially just Power Point slides; this is a good review if you attended the seminar, but without accompanying discussion it is unlikely a reader will learn retrofit methods by reading the materials alone
    - Contains good examples of some common mis-applications of hardware to avoid
    - Good start showing basic installation procedures for simple retrofits
7. **FEMA P-547** *Techniques for the Seismic Rehabilitation of Existing Buildings*—2006
    - An expansion of prior FEMA documents
    - Chapter 5 applies to residential construction; download from FEMA at: http://www.fema.gov/library/viewRecord.do?id=2393
    - First guideline I have seen that includes balloon framing retrofit details
8. **ABAG**—*Plan Set A*
    - Gives specific lengths of shear panels to install, based on building size and existing construction materials.
    - Limited to cripple-walls less than 4 feet tall.
    - FEMA *"South Napa Earthquake Recovery Advisory FEMA DR-4193-RA2"* (see Item 18 below) is meant to expand and improve upon Plan Set A.
9. **City of Seattle**—*Project Impact*
    - Similar to Plan Set A, except that instead of specific shear panel lengths it requires a certain percentage of wall length must receive shear panels, rather than absolute length (applies best to square buildings)
10. **LA**—*Standard Plan One*
    - Similar to Plan Set A, except that instead of specific shear panel lengths it requires a certain percentage of wall length must receive shear panels, rather than absolute length (applies best to square buildings). Includes conditions commonly found in the greater Los Angeles area.
11. **SEAOSC Seminar, 1994**—*"All About Wood"*
    - Detailed reporting of failures in the Northridge Earthquake
    - Recommendations for code changes (many of which have been adopted)

12. *California Existing Building Code* and *International Existing Building Code*
    - Essentially identical to LA and Seattle requirements;
    - Limited to 4-foot cripple walls, does not give absolute length requirements for shear panels
    - Basic construction details (about 4 different conditions with variations)
13. *Peace of Mind in Earthquake Country*—Peter Yanev, 2009 (Updated from the original 1974 edition, which, while it contained state-of-the-art information at the time, has been improved upon)
    - Excellent overview of geologic considerations, behavior of various high-rise and commercial building types, residential hazards (such as hillside homes) to look for.
    - Introduces cripple-wall retrofitting but the general nature of the book does not allow in-depth coverage of methods
14. **USGS—*Putting Down Roots in Earthquake Country***
    - Overview of seismology, shaking maps, fault locations
    - General homeowner tips for earthquake preparation and safety after a quake
15. **FEMA 50-1** *Seismic Rehabilitation Guidelines for Detached, Single-Family Wood-Framed Buildings* A compilation of information from prior FEMA publications, focusing on wood-framed construction
16. **City of San Jose—Apartment Owners Guide to Earthquake Safety**
17. *"Cyclic Tests of Engineered Shear Walls with Different Bottom Plate and Anchor Bolt Washer Sizes (Phase II)"* by Rakesh Gupta, Heather Redler, and Milo Clauson, Oregon State University, 2007
    www.awc.org/pdf/OSUFullScaleShearwallTestReportII-07.pdf
18. **FEMA "South Napa Earthquake Recovery Advisory FEMA DR-4193-RA2"** (Tentative title: "**Earthquake Strengthening of Cripple Walls in Wood-Frame Dwellings**") Currently under development. The plan set that accompanies this document expands on Plan Set A and includes retrofit requirements for cripple walls up to 7 feet tall. Expected to be released in June, 2015
19. *"Seismic Behavior of Level and Stepped Cripple Walls"* CUREE Publication W-17, Consortium of Universities for Research in Earthquake Engineering, 2002.
20. *"Seismic Behavior of Base-Level Diaphragm Anchorage of Hillside Woodframe Buildings,"* CUREE Publication W-24, Consortium of Universities for Research in Earthquake Engineering, 2003.

www.ingramcontent.com/pod-product-compliance
Lightning Source LLC
Chambersburg PA
CBHW081353290426
44110CB00018B/2364